Wireless Networks

Series Editor

Xuemin Sherman Shen, University of Waterloo, Waterloo, Canada

The purpose of Springer's Wireless Networks book series is to establish the state of the art and set the course for future research and development in wireless communication networks. The scope of this series includes not only all aspects of wireless networks (including cellular networks, WiFi, sensor networks, and vehicular networks), but related areas such as cloud computing and big data. The series serves as a central source of references for wireless networks research and development. It aims to publish thorough and cohesive overviews on specific topics in wireless networks, as well as works that are larger in scope than survey articles and that contain more detailed background information. The series also provides coverage of advanced and timely topics worthy of monographs, contributed volumes, textbooks and handbooks.

Shimin Gong · Dusit Niyato · Bo Gu ·
Kaibin Huang

Optimization-Driven Deep Reinforcement Learning for Wireless Networks

Springer

Shimin Gong
School of Intelligent Systems Engineering
Sun Yat-sen University
Guangzhou, China

Bo Gu
School of Intelligent Systems Engineering
Sun Yat-sen University
Guangzhou, China

Dusit Niyato
School of Computer Science
and Engineering
Nanyang Technological University
Singapore, Singapore

Kaibin Huang
Department of Electrical and Computer
Engineering
University of Hong Kong
Hong Kong SAR, China

ISSN 2366-1186 ISSN 2366-1445 (electronic)
Wireless Networks
ISBN 978-3-032-22996-0 ISBN 978-3-032-22997-7 (eBook)
https://doi.org/10.1007/978-3-032-22997-7

This Springer imprint is published by the registered company Springer Nature Switzerland AG
The registered company address is: Gewerbestrasse 11, 6330 Cham, Switzerland

Preface

Efficient resource utilization in wireless network is the key technology to accommodate the explosive growth of wireless users and the diverse service requirements. Conventionally it requires exact modeling of the wireless network and amenable problem formulation for optimal network design and resource allocation in an efficient and centralized manner. Such a model-based optimization strategy nowadays faces the computational challenges as the future wireless networks become more diversified, decentralized, dynamic, and autonomous. Instead of the conventional optimization methods, the model-free reinforcement learning (RL) can be a flexible framework for the wireless network to adapt its resource allocation based on continuous interactions among different network entities. The recent advent of deep reinforcement learning (DRL) has been shown its success to effectively address various network control problems, including channel access, relay selection, transmit beamforming, and user association, that can meet different performance improvement goals, in terms of capacity enhancement, coverage extension, energy- and spectrum-efficiency maximization. However, DRL still has obvious limitations for practical deployment due to insufficient training data or excessive learning overhead, especially for a large-scale decentralized wireless network. To facilitate resource-efficient design and practical deployment of the future autonomous wireless networks, more efficient learning structure and in-depth customization of DRL will be indispensable. It can be particularly important to integrate empirical knowledge or partial model information to improve the agility of autonomous decision-making in future wireless network.

This book intends to provide academia and industry with comprehensive coverage of the interplay between conventional model-based optimization and the model-free DRL methods in future wireless networks, focusing on various techniques for implementing and accelerating the learning performance in practice. Several essential contents will be elaborated in this book, including optimization-driven DRL that relies on model-based optimization to accelerate the model-free DRL, hierarchical DRL for more efficient learning in complex wireless networks, multi-agent DRL (MADRL) for large-scale decentralized wireless networks, Bayesian-enhanced DRL for handling incomplete information, and Lyapunov-guided DRL for dynamic resource allocation in wireless networks. Each technique is illustrated by case study in wireless networks, accompanied by in-depth discussions and insightful analysis.

The increasing heterogeneity and decentralized nature of future wireless networks create a complex environment for resource management, including terrestrial and aerial users, edge devices, and remote servers. Some nodes such as the network controllers may have greater capabilities for information collection, processing, and computation, allowing them to make more informed estimations and decisions. This motivates the optimization-driven DRL methods, which leverage partial information from various network entities to construct approximate optimization problems efficiently tractable to guide DRL toward a better learning result. This may significantly improve DRL's learning efficiency and reward performance. For large-scale wireless networks, hierarchical decomposition can be indispensable to achieve fast convergence and reliable decision-making. This motivates the hierarchical DRL framework that decomposes the control problem into different subproblems with different control variables. One part of the control variables can be iteratively updated by model-free DRL methods in the outer loop, while the other control variables can be efficiently determined by model-based optimization methods in the inner loop. Furthermore, a practical wireless network is inherently a multi-agent system with user devices of diverse computational capabilities, and thus it calls for efficient MADRL algorithms for coordinated control. It remains a challenging problem to speed up multi-agent learning with limited information exchange between numerous user devices in decentralized wireless networks. In dynamic wireless networks, complete network information is not always available due to time-varying channel conditions, mobility, or measurement delays. Data-driven prediction techniques and Bayesian optimization can be leveraged to mitigate such information loss, which motivates our design for Bayesian-enhanced DRL for more robust learning by leveraging Bayesian inference. For long-term performance guarantee in dynamic wireless networks, the optimal network control is commonly formulated into a stochastic optimization problem, which is computational demanding by conventional model-based optimization methods as resource demands and network conditions fluctuate randomly over time. This motivates the Lyapunov-guided DRL method, which first decomposes the multi-stage stochastic optimization into per-slot subproblems via Lyapunov optimization framework and then uses DRL method to adaptively solve the per-slot subproblem in each time slot based on current network states.

Finally, this book will conclude with promising research directions, hoping to stimulate future research in the field of DRL-driven intelligent wireless networks from different perspectives. The book is primarily tailored for professionals and researchers in wireless communications community focusing on artificial intelligence (AI)-driven network optimization. It also offers referential value to advanced-level undergraduate or graduate students majoring in electrical engineering or computer science. This book is the result of collaborative efforts and intellectual contributions from many individuals, and we are deeply grateful to all who have supported this endeavor. First, we extend our sincere appreciation to the co-authors of the case studies in different chapters, whose expertise, dedication, and insightful discussions were instrumental in shaping the content. Their collaborative spirit and rigorous research formed the foundation of this work. We also gratefully acknowledge the valuable contributions of several graduate students, whose active participation in literature review, simulation validation, figure preparation, and proofreading significantly enhanced the quality of the manuscript.

Guangzhou, China Shimin Gong
Singapore Dusit Niyato
Guangzhou, China Bo Gu
Hong Kong SAR, China Kaibin Huang

Competing Interests The authors have no competing interests to declare that are relevant to the content of this manuscript.

Contents

Acronyms

AF	Amplify-and-forward
AI	Artificial intelligence
AO	Alternating optimization
AoI	Age of information
AP	Access point
BCD	Block coordinate descent
BER	Bit-error-rate
BS	Base station
CGAN	Conditional generative adversarial network
CNN	Convolutional neural network
CSI	Channel state information
D2D	Device-to-device
DDPG	Deep deterministic policy gradient
DDQN	Double DQN
DF	Decode-and-forward
DNN	Deep neural network
DPG	Deterministic policy gradient
DQN	Deep Q-Network
DRL	Deep reinforcement learning
EXT	Extraction-then-Transmission
G2U	GU-to-UAV
GNN	Graph neural network
GU	Ground user
HAP	Hybrid access point
IRS	Intelligent reflecting surface
KL	Kullback–Leibler
LMI	Linear matrix inequality
LoS	Line-of-sight
MADDPG	Multi-agent DDPG
MADOL	Multi-agent deep option learning
MADRL	Multi-agent deep reinforcement learning

MAF	Max-Age-First
MDP	Markov decision process
MEC	Mobile edge computing
MISO	Multi-input single-output
ML	Machine learning
mmWave	Millimeter-wave
MRC	Maximal ratio combining
NLoS	Non-line-of-sight
PER	Prioritized experience replay
PPO	Proximal policy optimization
PS	Power-splitting
QoS	Quality of service
RF	Radio frequency
ROR	Round-Robin
SAoI	Semantic-aware AoI
SCA	Successive convex approximation
SDP	Semidefinite program
SDR	Semidefinite relaxation
SINR	Signal to interference plus noise ratio
SLAM	Simultaneous learning and modeling
SNR	Signal-to-noise ratio
SWIPT	Simultaneous wireless information and power transfer
TD	Temporal difference
TDMA	Time division multiple access
TRPO	Trust region policy optimization
TS	Time-switching
U2B	UAV-to-BS
U2U	UAV-to-UAV
UAV	Unmanned aerial vehicle
V2V	Vehicle-to-vehicle
VEC	Vehicular edge computing
WET	Wireless energy transfer
WPCN	Wireless-powered communication network

Chapter 1
Introduction

Modern wireless networks, such as the Internet of Things (IoT), mobile edge computing (MEC), and UAV-assisted wireless networks, are becoming increasingly complex, decentralized, and heterogeneous [1–5]. They involve a diverse array of network entities, including network controllers, mobile users, and vehicular devices, each with distinct capabilities, resource constraints, and service requirements. Due to strong spatio-temporal couplings among these devices, fundamental network control problems, such as channel access, data rate adaptation, power control, multi-antenna beamforming, and base station association, have grown significantly in complexity and dimensionality, rendering traditional optimization solutions impractical. Typically, optimal network control can be formulated as either a constrained optimization problem or a sequential decision-making process, aiming to achieve various design objectives, such as maximizing throughput or spectral efficiency, minimizing energy consumption, or reducing latency and age of information (AoI). However, as future wireless networks evolve toward large-scale and highly dynamic, they face unprecedented challenges, including imperfect modeling, uncertain system dynamics, and high-dimensional variable spaces. These factors make conventional approaches, such as convex optimization, dynamic programming, and stochastic optimization, increasingly inflexible, computationally prohibitive, or infeasible for real-time deployment.

As a promising breakthrough, deep reinforcement learning (DRL) has revolutionized the search for optimal control strategies in dynamic wireless networks [2, 6–8]. DRL overcomes the limitations of conventional reinforcement learning (RL) by integrating deep neural networks (DNNs), empowering autonomous and sequential decision making to solve high-dimensional and non-convex network control problems. Without knowing the complete network information, network entities can make local observations of the network environment and update the policies iteratively. By continuously interacting with the network environment, different network entities can build knowledge from experiences and learn optimal strategies for network control problems such as channel access, transmission scheduling, and resource allocation, to improve the overall network performance. In the literature, DRL has already been

S. Gong et al., *Optimization-Driven Deep Reinforcement Learning for Wireless Networks*, Wireless Networks, https://doi.org/10.1007/978-3-032-22997-7_1

shown its great success to effectively address various network control problems, in terms of capacity enhancement, coverage extension, energy- and spectrum-efficiency maximization, especially efficient for problems with exceptionally huge state and action spaces. DRL's flexibility and robustness make it highly attractive for large-scale and decentralized wireless networks, leading to a rapid expansion in various application scenarios and demonstrating its potential to revolutionize the network control paradigm.

In the following section, we first present a brief overview of RL/DRL fundamentals and analyze representative DRL algorithms, then we illustrate their applications and generic frameworks in different network control problems. To stimulate deeper discussion and motivate our case studies in the following chapters, we reveal potential interplays between conventional model-based optimization and model-free DRL, and their integrated designs for performance improvement in future wireless networks.

1.1 Fundamentals of RL and DRL

Machine learning (ML) approaches are typically categorized into three main paradigms: supervised learning, unsupervised learning, and RL. The key distinction between supervised and unsupervised learning lies in the availability of labeled sample data, i.e., supervised learning methods require labeled training samples to learn input-output mappings, while unsupervised learning methods aim to discover hidden patterns or structures in unlabeled data. In both cases, learning performance can be significantly enhanced by leveraging DNNs to extract high-level and nonlinear hidden features from training datasets. Different from supervised and unsupervised learning, RL focuses on sequential decision-making through continuous interaction between the RL agent and an uncertain environment [9]. This process is commonly formalized using a Markov decision process (MDP), properly defined by a set of system states, actions, transition matrix, and the reward function that quantifies the immediate reward for each action taken in each state. DRL extends traditional RL by employing DNNs to approximate RL's key components, such as policy functions, value functions, or even the reward and state transition models, enabling scalable and efficient learning in high-dimensional and complex environments. This integration enhances RL's flexibility, robustness, and generalization capability, making DRL particularly well-suited for dynamic control problems in wireless networks. In this section, we first review the fundamentals of RL and its extension to DRL, followed by a discussion of advanced techniques for improving the efficiency, stability, and convergence of DRL methods in practical network optimization problems.

RL enables an agent to make sequential decisions by interacting with an environment, i.e., it takes actions, observes the state transitions and rewards, and gradually updates its actions to converge toward an optimal policy. However, this learning process is typically time-consuming, as the agent has to extensively explore the environment and accumulate sufficient experience to adapt to the system dynamics. In large-scale and dynamic wireless networks, this exploratory phase becomes prohibitively slow with huge state and action spaces, leading to poor sample efficiency and delayed convergence. As a result, traditional RL methods often fail to be applicable in modern wireless systems. As deep learning plays a pivotal role in advancing artificial intelligence (AI), its capabilities of modeling high-dimensional nonlinear relationships can be leveraged to overcome the limitations of RL, and thus motivate the recent development of more efficient and scalable DRL framework to accelerate learning and improve adaptability in complex environments. DRL integrates DNNs into the RL framework to approximate value functions, policies, or transition models. The DNNs' expressive power enables DRL to handle high-dimensional state and action spaces that are intractable for traditional RL methods. As a result, DRL has been successfully applied to a wide range of complex domains, including robotics, control systems, wireless networks, computer vision, speech recognition, and natural language processing.

RL provides an effective solution framework to MDP, which is characterized by a state space $\mathcal{S}$, an action space $\mathcal{A}$ to interact with the environment, and reward function $r(\boldsymbol{s}_t, \boldsymbol{a}_t)$ to evaluate the immediate feedback for taking action $\boldsymbol{a}_t \in \mathcal{A}$ in state $\boldsymbol{s}_t \in \mathcal{S}$. In each time step t, the RL agent takes an action $\boldsymbol{a}_t$ based on an observation of the system state $\boldsymbol{s}_t$ and receives an immediate reward, quantifying the desirability of the state-action pair $(\boldsymbol{s}_t, \boldsymbol{a}_t)$. The immediate reward $r(\boldsymbol{s}_t, \boldsymbol{a}_t)$ guides the RL agent to iteratively adapt its decision-making strategy to maximize the cumulative discounted reward:

$$V^{\pi}(\boldsymbol{s}_t) = \mathbb{E}\Big[\sum_{i=0}^{\infty} \gamma^i r(\boldsymbol{s}_{t+i}, \boldsymbol{a}_{t+i})\Big],$$

where constant $\gamma \in (0, 1)$ represents the discount factor to balance the importance of immediate and future rewards. The action $\boldsymbol{a}_t$ executed in the environment with the current state $\boldsymbol{s}_t$ may cause the state transition to $\boldsymbol{s}_{t+1}$, according to the underlying transition dynamics. The immediate reward and the state transition will guide the adaptation of the agent's decision-making strategy, known as the policy π specifying a sequence of actions the agent should take as the system evolves in different states.

The cumulative reward $V^{\pi}(\boldsymbol{s}_t)$ is also known as the state-value function, which determines the expected long-term return starting from the current state $\boldsymbol{s}_t$ and thereafter following the same policy π. The learning process continues iteratively as an optimal policy can be found to maximize the expected cumulative reward. The maximization objective can be formally characterized by using either the state-value function $V^{\pi}(\boldsymbol{s}_t)$ or the action-value function defined as follows:

$$\begin{aligned} Q^{\pi}(\boldsymbol{s}_t, \boldsymbol{a}_t) &= r(\mathbf{s}_t, \mathbf{a}_t) + \gamma \sum_{\boldsymbol{s}_{t+1} \in \mathcal{S}} p(\boldsymbol{s}_{t+1} | \boldsymbol{s}_t, \boldsymbol{a}_t) V^{\pi}(\boldsymbol{s}_{t+1}) \\ &= r(\mathbf{s}_t, \mathbf{a}_t) + \gamma \mathbb{E}[V^{\pi}(\boldsymbol{s}_{t+1})]. \end{aligned}$$

The probability $p(\boldsymbol{s}_{t+1} | \boldsymbol{s}_t, \boldsymbol{a}_t)$ determines the probability of system transition to the next state $\boldsymbol{s}_{t+1}$. The action-value function $Q^{\pi}(\boldsymbol{s}_t, \boldsymbol{a}_t)$ is also known as the Q-value, mapping each state-action pair to a cumulative future reward. RL provides a solution framework to search for the optimal policy π^* maximizing the state-value $V^*(s_t) = \max_\pi V^\pi(s_t)$. By leveraging Bellman equation [9], we can transform the policy optimization into a series of action optimizations at different states:

$$V^*(\mathbf{s}_t) = \max_{\mathbf{a}_t \in \mathcal{A}} r(\mathbf{s}_t, \mathbf{a}_t) + \gamma \mathbb{E}\left[V^*(\mathbf{s}_{t+1})\right] = \max_{\mathbf{a}_t \in \mathcal{A}} Q^*(s_t, a_t).$$

This transformation lays the foundation for dynamic programming and iterative RL algorithms such as Q-learning and policy gradient methods.

The well-established Q-learning algorithm performs effectively in environments with small and finite state and action spaces, by iteratively updating and maximizing the Q-value in each state, i.e., $\mathbf{a}_t^* = \arg\max_{\mathbf{a}_t \in \mathcal{A}} Q^*(\mathbf{s}_t, \mathbf{a}_t)$. For algorithmic implementation, the Q-value can be randomly initialized and then iteratively updated using the mean squared error (MSE) between the current Q-value and its target value y_t estimated by looking one step ahead:

$$Q_{t+1}(\mathbf{s}_t, \mathbf{a}_t) = Q_t(\mathbf{s}_t, \mathbf{a}_t) + \tau_t \Big[y_t - Q_t(\mathbf{s}_t, \mathbf{a}_t)\Big], \tag{1.1}$$

where τ_t is a small step-size (or learning rate) controlling the weight of the update to Q-value. The target y_t is evaluated on the next state as follows:

$$y_t = r(\mathbf{s}_t, \mathbf{a}_t) + \gamma \max_{\mathbf{a}} Q_t(\mathbf{s}_{t+1}, \mathbf{a}).$$

The difference $y_t - Q_t(\mathbf{s}_t, \mathbf{a}_t)$ is also termed as the temporal-difference (TD) error. In each time slot t, a small part of the TD error is updated to the Q-value. This enables the agent to gradually improve its Q-value estimates toward the optimal action-value function.

RL methods are primarily categorized into value-based and policy-based approaches [9, 10]. The value-based approach focuses on estimating the value function (either state-value or action-value function) and selects actions to maximize this estimated value directly in an iterative process. Classic methods include value iteration, which evaluates the optimal value function by iteratively applying the Bellman optimality equation, and the Q-learning algorithm, which directly learns the optimal action-value function through TD updates as shown in (1.1). The policy-based approach parametrizes the policy function and improves the value function by directly updating the policy parameter in gradient-based methods. RL can also be categorized into on- and off-policy methods based on their data usage. The on-policy learning requires that the experience used for training is generated by the current policy, i.e., all actions during exploration and evaluation are chosen according to the same policy. While on-policy ensures consistency between the behavior and target policies, it results in low sample efficiency and thus requires more interactions with the environment to collect sufficient training samples for policy updates. This makes it computationally expensive and less practical for solving complicated real-world problems. In contrast, off-policy learning allows the reuse of historical experience collected under previous or different policies, significantly improving sample efficiency. However, this flexibility can introduce training instability due to distributional shift and bootstrapping errors, necessitating careful hyperparameter tuning to ensure stable and convergent learning. This motivates the development of various techniques such as fixed target networks, experience replay, and reward clipping to mitigate variance in policy updates.

1.1.1 Deep Q-Network (DQN)

RL becomes unstable and may fail to converge when it is applied to complex environments with large or continuous state and action spaces. Primarily, DRL overcomes this difficulty by using DNNs as parameterized function approximators for value function and policy. A landmark in this direction is the success of deep Q-network (DQN) algorithm [11], which relies on the use of DNNs to approximate the action-value function $Q(\boldsymbol{s}_t, \boldsymbol{a}_t)$. Deep deterministic policy gradient (DDPG) algorithm proposed in [12] adopts an actor-critic framework using two separate DNNs to estimate the value function and the policy function, respectively. This decoupled structure enables efficient learning in continuous action spaces and has inspired numerous advanced DRL algorithms to improve stability, sample efficiency, and convergence in complex environments.

DQN extends the classical Q-learning algorithm by employing DNNs as a parametric function approximator for the Q-value. For control problems with small-size state and action spaces, Q-learning maintains a lookup table to store and update the Q-values for each state-action pair. However, this tabular approach becomes infeasible in high-dimensional or continuous spaces. DQN overcomes this limitation by approximating the Q-value as $Q(\mathbf{s}_t, \mathbf{a}_t|\boldsymbol{\omega}_t)$, also known as the Q-network, where $\boldsymbol{\omega}_t$ denotes the trainable DNN weight parameters. This makes it flexible for complex control problems with very large state and action spaces. As illustrated in Fig. 1.1a, the Q-network takes the current state $\mathbf{s}_t$ as input and outputs the estimated Q-values for different actions, enabling scalable and generalizable Q-value estimation from high-dimensional inputs.

When generating the action, DQN uses the ϵ-greedy policy for action selection and uses the same Q-network to evaluate the target Q-value in TD updates. This duplicated use of the same network for both action selection and evaluation may lead to over-optimistic Q-value estimates. To correct this, Double DQN (DDQN) [13] decouples action selection from Q-value evaluation by using two sets of DNNs, i.e., it relies on the online Q-network to select the best action while the target Q-network provides the Q-value evaluation for the selected action, as illustrated in Fig. 1.1b. This modification can reduce maximization bias and improve the accuracy of Q-value estimation. Dueling DQN [14] is another DQN variant that decomposes the Q-value into two separate streams, i.e., the state-value and the advantage-value, approximated by two independent DNNs with a shared neural network in a dueling architecture, as shown in Fig. 1.1c. The state-value represents the expected return from a given state regardless of the action, while the advantage-value evaluates the relative benefit of taking a specific action over the other actions in a given state. Two parallel streams are then combined via an aggregating layer to produce the Q-value estimates.

The success of DQN and its variants depends on two key mechanisms [11], i.e., experience replay decorrelates training data and the target Q-network provides stable learning targets. They work in tandem to stabilize learning with large state and action spaces. The experience replay stores historical transition samples in a replay memory buffer and randomly selects a subset of samples, that is, a minibatch, from the replay memory to update DNNs. This random minibatch breaks the temporal correlations between consecutive training samples, and ensures more efficient and stable learning by reusing the experience. To further improve sample efficiency, prioritized experience replay (PER) strategy can be leveraged to assign higher sampling probabilities to the transition samples that potentially generate higher rewards [15]. A common and effective PER strategy is to prioritize transition samples by the magnitude of their TD-errors, which reflects the discrepancy between the Q-value estimate and its target. Transition samples with larger TD-errors indicate greater prediction inaccuracies and thus offer higher potential for improving the Q-value estimates. Practically, the TD-error based PER is often combined with random sampling to ensure sufficient exploration of the entire experience space and maintain robust convergence.

The target Q-network is used to ensure a smooth training of the online Q-network and implemented as an delayed copy of the online Q-network. Given the DNN parameters $\boldsymbol{\omega}_t$ and $\boldsymbol{\omega}_t'$ for the online and target Q-networks, respectively, their corresponding outputs are $Q(\mathbf{s}_t, \mathbf{a}_t|\boldsymbol{\omega}_t)$ and $Q(\mathbf{s}_t, \mathbf{a}_t|\boldsymbol{\omega}_t')$. At each time step t, the DQN agent observes the current system state $\mathbf{s}_t$ and chooses the action $\mathbf{a}_t$ to maximize the online Q-value $Q(\mathbf{s}_t, \mathbf{a}_t|\boldsymbol{\omega}_t)$. Upon executing the action $\mathbf{a}_t$ in the environment, the agent receives the immediate reward $r_t(\mathbf{s}_t, \mathbf{a}_t)$ and transitions to the next state $\mathbf{s}_{t+1}$. The transition sample $(\mathbf{s}_t, \mathbf{a}_t, v_t, \mathbf{s}_{t+1})$ is then stored in the experience replay buffer for later use in training. For each sample in the minibatch, the DQN agent estimates the target Q-value by looking one step ahead using the target Q-network with DNN parameter $\boldsymbol{\omega}_t'$, expressed as follows:

$$y_t = v_t(\mathbf{s}_t, \mathbf{a}_t) + \gamma Q(\mathbf{s}_{t+1}, \mathbf{a}_{t+1}|\boldsymbol{\omega}_t'). \tag{1.2}$$

The target Q-network updates its parameters $\boldsymbol{\omega}_t'$ by periodically copying the weights from the online Q-network $\boldsymbol{\omega}_t$, as illustrated in Fig. 1.1. Such a delayed copying of DNN parameters can provide consistent target Q-values and thus help stabilize the learning process. The experience replay provides the mini-batch of historical transition samples for training, while the target Q-network sets the training target as shown in (1.2). The DNN training in each step aims at adapting the DNN parameter $\boldsymbol{\omega}_t$ to minimize a loss function $\ell(\boldsymbol{\omega}_t)$, defined in terms of the TD-error and given as follows:

$$\ell(\boldsymbol{\omega}_t) = \mathbb{E}_{\mathcal{B}_t}[|y_i - Q(\mathbf{s}_i, \mathbf{a}_i|\boldsymbol{\omega}_t)|^2]. \tag{1.3}$$

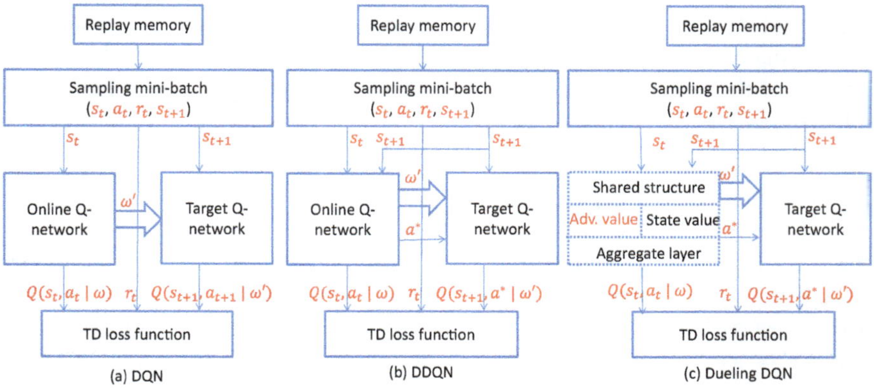

Fig. 1.1 The comparison of DQN and its variants [10]

The expectation in (1.3) is approximated by averaging over the mini-batch $\mathcal{B}_t$ to compute the gradient and update the DNN parameters at each training iteration. For each transition sample $i \in \mathcal{B}_t$, we can generate the online Q-value $Q(\mathbf{s}_i, \mathbf{a}_i|\boldsymbol{\omega}_t)$ and estimate the target value y_i in (1.2). By minimizing the loss $\ell(\boldsymbol{\omega}_t)$, the DNN parameters $\boldsymbol{\omega}_t$ are updated through backpropagation of the gradient information.

1.1.2 Deep Deterministic Policy Gradient (DDPG)

Unlike the value-based DQN method, which indirectly derives a policy by learning and maximizing the action-value function, the policy-based approach directly optimizes the parametric policy π_ω in gradient-based methods to maximize the expected cumulative reward. The policy-based and value-based approaches can be effectively combined within the actor-critic framework [9]. The critic evaluates the current policy by estimating the action-value or state-value function, typically by minimizing the TD-error to produce a low-variance value estimation. The actor then uses this value evaluation as a feedback signal to update the policy parameters by policy gradient methods. To enhance scalability and generalization, two separate DNNs are used as the parametric approximations for the critic and actor, respectively. Let $\pi_{\boldsymbol{v}}(\mathbf{a}|\mathbf{s})$ denote the DNN parametric actor network. It outputs a probability distribution over different actions $\mathbf{a} \in \mathcal{A}$ for discrete action spaces, or the parameters of a stochastic policy for continuous control tasks. The DNN training for the actor network aims to update the parameters $\boldsymbol{v}$ by maximizing the expected cumulative reward, reformulated as follows:

$$J(\boldsymbol{v}) = \sum_{\mathbf{s} \in \mathcal{S}} d(\mathbf{s}) V_\pi(\mathbf{s}) = \sum_{\mathbf{s} \in \mathcal{S}} d(\mathbf{s}) \sum_{\mathbf{a} \in \mathcal{A}} \pi_{\boldsymbol{v}}(\mathbf{a}|\mathbf{s}) Q^\pi(\mathbf{s}, \mathbf{a}), \tag{1.4}$$

where $d(\mathbf{s})$ denotes the stationary distribution over system states and $Q^\pi(\mathbf{s}, \mathbf{a})$ is the Q-value obtained from the policy $\pi_{\boldsymbol{v}}$. The value function $J(\boldsymbol{v})$ now becomes a function of the policy parameters $\boldsymbol{v}$. According to the policy gradient theorem [9, 12], the gradient of $J(\boldsymbol{v})$ can be expressed in an analytical form:

$$\nabla_{\boldsymbol{v}} J(\boldsymbol{v}) = \mathbb{E}_\pi\Big[Q^\pi(\mathbf{s}, \mathbf{a})\nabla_{\boldsymbol{v}} \ln \pi_{\boldsymbol{v}}(\mathbf{a}|\mathbf{s})\Big], \tag{1.5}$$

where the expectation is taken over state-action pairs collected under the policy $\pi_{\boldsymbol{v}}$ and the system's transition dynamics. As shown in (1.5), the policy gradient theorem establishes a direct link between the policy gradient $\nabla_{\boldsymbol{v}} J(\boldsymbol{v})$ and the Q-value $Q^\pi(\mathbf{s}, \mathbf{a})$. Specifically, it decomposes the gradient into the evaluation of the Q-value and the gradient of the log-policy $\nabla_{\boldsymbol{v}} \ln \pi_{\boldsymbol{v}}(\mathbf{a}|\mathbf{s})$, averaged over the state and action spaces under the current policy.

For practical implementation, the policy gradient $\nabla_{\boldsymbol{v}} J(\boldsymbol{v})$ can be evaluated by sampling transition trajectories. At the current state $\mathbf{s}_t$, the agent selects an action $\mathbf{a}_t$ according to the policy $\pi_{\boldsymbol{v}}$, executes it in the environment, collects an immediate reward r_t and observes the transition to the next state $\mathbf{s}_{t+1}$, which constitutes a transition trajectory as the agent continuously interacts with the environment. Let $g_t = \sum_{\tau=t}^{T} \gamma^{\tau-t} r_\tau$ denote the sample estimate for the Q-value following the transition trajectory. We can approximate the policy gradient $\nabla_{\boldsymbol{v}} J(\boldsymbol{v})$ by the Monte Carlo estimate $g_t \nabla_{\boldsymbol{v}} \ln \pi_{\boldsymbol{v}}(\mathbf{a}_t|\mathbf{s}_t)$ and update the policy network in a gradient ascend direction as follows:

$$\boldsymbol{v} \leftarrow \boldsymbol{v} + \alpha_{\boldsymbol{v}} g_t \nabla_{\boldsymbol{v}} \ln \pi_{\boldsymbol{v}}(\mathbf{a}_t|\mathbf{s}_t),$$

where $\alpha_{\boldsymbol{v}}$ is a small step-size for gradient update. In addition to using the Monte Carlo estimation g_t based on transition trajectories, we can employ a separate DNN with parameters $\boldsymbol{\omega}$ to approximate the Q-value $Q^\pi(\mathbf{s}, \mathbf{a})$ in (1.5), similar to that in the DQN algorithm. This leads to the actor-critic framework that integrates the policy-based and value-based approaches. The actor updates the policy using policy gradient methods based on the critic's policy evaluation, while the critic improves the Q-value estimates by minimizing a loss function similar to (1.3). We can take the derivative of the loss function and update $\boldsymbol{\omega}$ in the gradient direction as follows:

$$\boldsymbol{\omega} \leftarrow \boldsymbol{\omega} + \alpha_\omega \delta_t \nabla_\omega Q(\mathbf{s}_t, \mathbf{a}_t|\boldsymbol{\omega}),$$

where $\delta_t = y_t - Q(\mathbf{s}_t, \mathbf{a}_t|\boldsymbol{\omega}')$ is the TD-error and α_ω denotes a small step-size. The Q-value estimate can also be replaced by the advantage function to reduce the variance of the policy gradient estimates, leading to more stable and efficient learning.

The policy gradient theorem can be extended to the deterministic policy gradient (DPG) framework [12], where the policy $\pi_{\boldsymbol{v}}(\mathbf{s}_t)$ outputs a deterministic action rather than a probability distribution over actions. The DPG framework eliminates the need to integrate over the action space, leading to lower-variance gradient estimates and improved sample efficiency. The DDPG algorithm is built on the DPG theorem and following the actor-critic framework by using DNN approximations for both Q-value

$Q(\mathbf{s}, \mathbf{a}|\boldsymbol{\omega})$ and the deterministic policy $\pi_{\boldsymbol{v}}(\mathbf{s}_t)$. Specifically, the DPG theorem helps simplify the policy gradient in (1.5) as follows:

$$\nabla_{\boldsymbol{v}} J(\boldsymbol{\omega}, \boldsymbol{v}) = \mathbb{E}_{\mathbf{s}\sim d(\mathbf{s})}[\nabla_{\mathbf{a}} Q(\mathbf{s}, \mathbf{a}|\boldsymbol{\omega}) \nabla_{\boldsymbol{v}} \pi_{\boldsymbol{v}}(\mathbf{s})|_{\mathbf{a}=\pi_{\boldsymbol{v}}(\mathbf{s})}].$$

As such, the actor network updates its parameters $\boldsymbol{v}$ by using the stochastic gradient ascent method as follows:

$$\boldsymbol{v} \leftarrow \boldsymbol{v} + \alpha_{\boldsymbol{v}} \nabla_{\mathbf{a}} Q(\mathbf{s}, \mathbf{a}|\boldsymbol{\omega}) \nabla_{\boldsymbol{v}} \pi_{\boldsymbol{v}}(\mathbf{s})|_{\mathbf{a}=\pi_{\boldsymbol{v}}(\mathbf{s})}.$$

Similarly, the critic network updates its parameters to minimize the mean squared TD-error as follows:

$$\boldsymbol{\omega} \leftarrow \boldsymbol{\omega} + \alpha_{\omega} \delta_t \nabla_{\omega} Q(\mathbf{s}, \mathbf{a}|\boldsymbol{\omega}).$$

For practical implementation, the critic's training is to minimize the loss function in (1.3), by sampling a minibatch from the experience replay memory, where the target value y_i of each training sample is estimated as follows:

$$y_i = r(\mathbf{s}_i, \mathbf{a}_i) + \gamma Q(\mathbf{s}_{i+1}, \pi(\mathbf{s}_{i+1}|\boldsymbol{v}')|\boldsymbol{\omega}').$$

Similar to DQN, DDPG stabilizes learning by maintaining target networks for both the actor and the critic. The DNN parameters of these target networks ($\boldsymbol{\omega}'$ for critic and $\boldsymbol{v}'_t$ for actor) are slowly updated copies of their online networks ($\boldsymbol{\omega}$ and $\boldsymbol{v}$), respectively.

1.1.3 Trust Region Policy Optimization (TRPO)

The DDPG algorithm fundamentally relies on policy gradient estimation for training its deterministic policy. However, even with accurate gradient estimates, the subsequent updates to the DNN parameter often exhibit instability as it is still challenging to select an appropriate gradient step size. Improper step sizes may lead to divergent parameter updates or excessively slow convergence, leading to bias into the value function estimation. This biased value estimation then propagates errors back into the policy gradient evaluation, creating a detrimental feedback loop and resulting in divergence of the training process. This challenge motivates the core design principle behind the TRPO algorithm [16], which introduces a trust region constraint for the gradient-based policy updates. Unlike unconstrained gradient methods, the trust region explicitly constrains the policy updates to prevent excessive deviation from the previous policy. The distance between two successive policies is typically defined in terms of Kullback-Leibler (KL) divergence. By limiting the policies' KL divergence, TRPO prevents overly aggressive updates that may degrade policy performance and thus achieves significantly more stable learning performance compared to conventional policy gradient approaches.

Before examining TRPO's technical details, we revisit the policy gradient in (1.5) and consider an important reformulation to enhance learning efficiency. It is clear that the gradient estimate in (1.5) requires the state-action trajectory generated by the current target policy π_v. Such a on-policy learning scheme prevents the reuse of past experiences, significantly limiting the sample efficiency. However, via importance sampling, we can derive an off-policy reformulation of the policy gradient in (1.5) as follows that permits the use of trajectories collected by previous policies.

$$\nabla_v J(v) = \mathbb{E}_{\pi_v^o}\Big[\frac{\pi_v(\mathbf{s}, \mathbf{a})}{\pi_v^o(\mathbf{s}, \mathbf{a})} A_{\pi_v}(\mathbf{s}, \mathbf{a}) \nabla_v \ln \pi_v(\mathbf{a}|\mathbf{s})\Big], \tag{1.6}$$

where π_v represents the target policy we need to optimize and π_v^o denotes the previous policy used for collecting the training samples, also called the behavior policy. The Q-value in (1.5) is replaced by the advantage function $A_{\pi_v}(\mathbf{s}, \mathbf{a})$ to reduce variability and enhance learning efficiency. The reformulation in (1.6) yields an off-policy gradient estimate by leveraging past experience collected under a different and potentially obsolete behavior policy π_v^o. This enables a significant improvement in sample efficiency through the use of an experience replay buffer, similar to the DQN algorithm.

To stabilize policy updates, TRPO imposes an additional constraint to enforce that the new updated policy π_v does not change excessively from the behavior policy π_v^o, leading to a constrained optimization as follows:

$$\max_{v} \; \mathbb{E}_{\pi_v^o}\Big[\frac{\pi_v(\mathbf{s}, \mathbf{a})}{\pi_v^o(\mathbf{s}, \mathbf{a})} A_{\pi_v^o}(\mathbf{s}, \mathbf{a})\Big], \tag{1.7a}$$

$$\text{s.t.} \; \mathbb{E}_{\pi_v^o}[D_{KL}(\pi_v, \pi_v^o)] \le \delta_{KL}, \tag{1.7b}$$

where $D_{KL}(\cdot, \cdot)$ denotes the KL divergence between two probability distribution functions [17]. The inequality constraint in (1.7b) sets an upper bound δ_{KL} on the KL divergence of two policies π_v and π_v^o. The advantage function $A_{\pi_v^o}$ in (1.7a) provides an approximation of the true advantage A_{π_v} of the target policy π_v. However, the optimization problem in (1.7) does not admit an efficiently tractable solution. The KL divergence introduces complex nonlinear couplings in the policy space. Normally, we require computationally expensive second-order expansion for the trust region constraint, making it complicated for gradient updates.

Proximal policy optimization (PPO) proposed in [18] provides a simple and efficient approximation to problem (1.7) by transforming it into a unconstrained optimization with a new surrogate objective function, which includes a penalty term to discourage excessively large policy updates. First, PPO revises the objective in (1.7) by limiting the importance weight $\rho_v \triangleq \frac{\pi_v(\mathbf{s},\mathbf{a})}{\pi_v^o(\mathbf{s},\mathbf{a})}$ within a clipped interval $[1-\epsilon, 1+\epsilon]$:

$$\tilde{J}(v) = \mathbb{E}_{\pi_v^o}\Big[\min\Big\{\rho_v A_{\pi_v^o}, \text{clip}(\rho_v, 1-\epsilon, 1+\epsilon) A_{\pi_v^o}\Big\}\Big],$$

where the clip function $\text{clip}(\rho_v, 1-\epsilon, 1+\epsilon)$ limits ρ_v to $[1-\epsilon, 1+\epsilon]$, and ϵ defines the clipping range. It returns $1-\epsilon$ if $\rho_v < 1-\epsilon$ and $1+\epsilon$ if $\rho_v > 1+\epsilon$. This clipping mechanism for the approximate value function $\tilde{J}(\boldsymbol{v})$ prevents large deviations of the target policy π_v from the behavior policy π_v^o, stabilizing policy updates for any advantage value $A_{\pi_v^o}$. Second, PPO introduces a Lagrangian dual variable β_{KL} and reformulates the trust region constraint (1.7b) into the unconstrained maximization problem as follows:

$$\max_{\boldsymbol{v}} \ \tilde{J}(\boldsymbol{v}) - \beta_{KL}\mathbb{E}_{\pi_v^o}\Big[D_{KL}(\pi_v, \pi_v^o)\Big]. \tag{1.8}$$

The policy parameters $\boldsymbol{v}$ for the new value function in (1.8) can be efficiently maximized using the stochastic gradient ascent method. Furthermore, the dual variable β_{KL} can be dynamically adjusted to ensure that the measured KL divergence $\mathbb{E}_{\pi_v^o}\Big[D_{KL}(\pi_v, \pi_v^o)\Big]$ is close to the target value δ_{KL}.

As shown in Fig. 1.2, we compare DQN and PPO in the context of wireless networks. Unlike DQN, PPO employs an actor-critic architecture utilizing two DNNs to approximate the policy and the Q-value, respectively. The behavior policy interacts with the environment, generating transition samples in the experience replay buffer used to train the target policy. The update of the target policy requires to maximize the objective in (1.8), which relies on an importance sampling ratio $\rho_v(\mathbf{s}, \mathbf{a})$ measuring the difference between the target policy π_v and the behavior policy π_v^o.

1.2 DRL Applications in Wireless Networks

As modern wireless networks become increasingly large-scale and complicated, network control problems involve highly heterogeneous decision spaces, encompassing

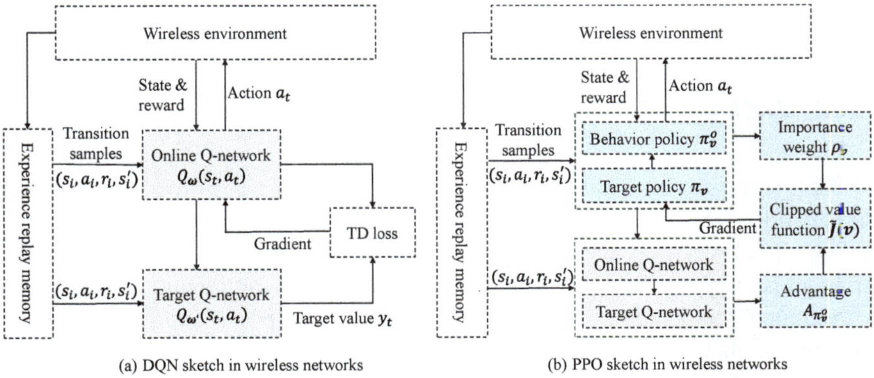

(a) DQN sketch in wireless networks (b) PPO sketch in wireless networks

Fig. 1.2 Comparison of DQN and PPO in wireless networks [19]

both discrete variables (e.g., user scheduling, mode selection, and on/off switching) and continuous variables (e.g., power allocation, beamforming, and resource allocation). This heterogeneity motivates the integration of value-based and policy-based DRL methods for mixed decision-making problems. In the following sections, we present representative applications of DRL in wireless networks, where agents deal with complex and dynamic environments, involving multi-agent interactions, partial observability, and coupled control across both discrete and continuous domains.

1.2.1 DRL for Decentralized Access Control

In decentralized and ad-hoc wireless networks, different network entities such as sensors and wireless users prefer to make independent control decisions, e.g., channel access, transmission control, and user associations, to achieve individuals' objectives, e.g., throughput maximization and energy minimization. However, this is practically challenging due to the highly dynamic and uncertain nature of wireless networks. DRL algorithms such as DQN and DDPG can learn and build knowledge about the network dynamics to make optimal decisions. DRL has been shown effective particularly for dynamic spectrum access [7, 20–22]. When wireless users have limited observations of the network environment, e.g., channel states, DRL has proven to be an effective tool for enabling them to dynamically and intelligently select optimal channels to maximize their long-term throughput based on local observations. The joint user association and spectrum access problem is inherently combinatorial and non-convex with high computational complexity. This make it computationally prohibitive for traditional optimization approaches, especially in large-scale dynamic and heterogeneous wireless networks. Similarly, for adaptive rate control, DRL can be adopted effectively for wireless users to choose different bitrates dynamically for more efficient data transmissions, instead of dynamic programming approaches which typically require complete system information and suffer from high computational complexity.

1.2.2 DRL for Mobile Edge Computing

By deploying computational resources, caching capabilities, and processing power at the network edge or closer to end users, MEC significantly enhances energy efficiency and quality of service (QoS) for applications with high computational demands and stringent latency requirements. This is particularly useful for the future IoT systems with limited computation resource, memory space, and power supplies to support emerging applications integrated with AI capabilities, e.g., semantic analysis, knowledge extraction, and inference [23, 24]. MEC enables edge users such as sensors, wearable devices, and handheld terminals to offload their data and computational workloads to nearby MEC servers or collaborative mobile users with complementary resources. Through intelligent allocation of computational resources and transmission control, such MEC offloading paradigm can significantly reduce end-to-end delay, extends battery lifetime, and potentially enhance security for latency-sensitive and computation-intensive IoT applications [25–27].

One of the key design challenges in MEC systems is the joint optimization of data offloading rates, workload allocation, and MEC server selection. This must account for time-varying channel conditions, user mobility, limited energy supplies, dynamic workloads, and heterogeneous resource constraints at both end-user devices and MEC servers. A joint optimization framework for caching, networking, computation allocation, and transmission control in MEC systems typically requires complex system analysis due to implicit and strong couplings among network entities, as well as heterogeneities in QoS requirements and wireless resources. Such an optimization framework is also inherently limited in its ability to capture uncertain network dynamics, such as time-varying channel conditions, unpredictable workloads, and fluctuating energy availability. DRL avoids above-mentioned difficulties by reformulating the MEC system's optimal control into a sequential decision making problem in the MDP framework. Without requiring prior knowledge of the network dynamics, DRL enables mobile users to progressively adapt offloading policies based on past experiences. Practically, DRL can be deployed at a central network controller, such as the MEC server, which is rich in computational resources and well-suited for information aggregation, system-wide coordination, and cross-layer policy design [7, 10, 25].

As illustrated in Fig. 1.3, the generic DRL framework is customized for MEC systems to optimize diverse performance objectives, comprising the following key components [10]:

- Network profiling: It helps extract problem-specific information closely relevant to the network control problems. The network environment can be complex, dynamic, and contains high-dimensional data. With limited local observations, data-driven or model-based prediction methods can be used to help estimate the system-wide information for decision making.
- State orchestration: This component aims to select the most salient and representative state variables, minimizing the dimensionality of the state space while preserving learning performance. It can be achieved through dimensionality reduction by focusing on problem-relevant principal components. Typically, the network performance depends on the resource demands and supplies. Hence, the system state can represent the dynamics of resource consumption and its regeneration.
- Training scheme: This component orchestrates the training of the value and policy networks to learn hybrid offloading decisions, encompassing discrete actions such as channel assignment and user association, as well as continuous variables including bandwidth allocation and beamforming. To improve learning efficiency, the model-free training can be integrated with the model-based optimization methods to avoid unnecessary action exploration. The control variables can be also divided into hierarchical learning loops to better suit the decentralized and heterogeneous network structure.

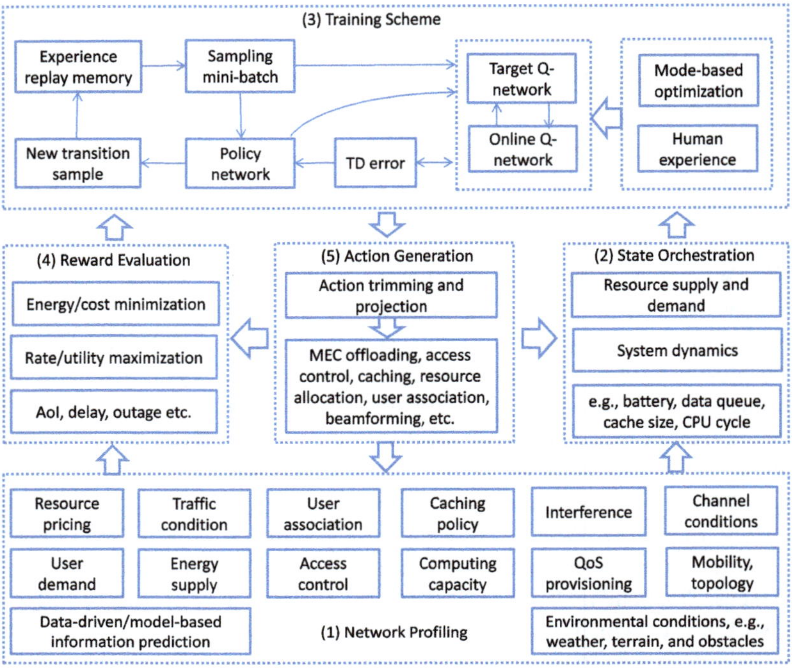

Fig. 1.3 Generic DRL framework for resource allocation in MEC systems [10]

- Reward evaluation: It enables the DRL agent to dynamically update its MEC offloading policy at every decision epoch based on the observed system state. In practice, the reward can be defined as the objective of the network control problem, such as long-term throughput or energy efficiency, and it is typically evaluated at the end of each decision epoch. To enable timely learning, a model-based predictor can be integrated to estimate the expected immediate reward by forecasting future system dynamics, such as workload evolution and channel variations.
- Action generation: This component maps the raw output of the DRL agent to a feasible action. For each system state, the agent produces an action vector, which is then projected onto valid system configurations through a feasibility check. This projection may involve quantization or approximation to convert continuous outputs into discrete control decisions, such as server activation or user association. Additionally, random noise can be injected into continuous actions (e.g., beamforming or bandwidth allocation) to enhance exploration during training.

The generic DRL framework can be customized to optimize MEC systems to meet diverse performance requirements, such as latency reduction, energy efficiency, or resource utilization. Depending on the system configuration and action space, different DRL algorithms are employed to address specific challenges. In scenarios with one user device and multiple access points (APs) or edge servers [28], DQN can be used to learn an optimal selection policy that minimizes the user's energy consumption and communication cost without prior knowledge of the user's mobility pattern. The offloading decision is based on predictions of the user's future location and remaining data size. Similarly, both RL and DQN were applied in [29] to optimize the edge server selection and offloading rate for an energy constrained IoT device with energy harvesting capability. When multiple users concurrently request computation resources from a shared MEC server [30], efficient spectrum and computational resource allocation becomes critical to maximize the sum computational rate or to minimize service delay and power consumption. In this setting, the DRL agent can learn both binary offloading decisions or continuous actions, e.g., bandwidth and computational resource allocation, under constraints such as limited server capacity and time-varying channel conditions. Considering a distributed infrastructure with multiple BSs and MEC servers, each user's computation offloading can be routed through different BSs to available MEC servers. To minimize end-to-end processing delay or a long-term cost, DQN and DDQN can be applied to jointly optimize user association and transmission control strategies, e.g., [31, 32]. The system state includes channel conditions between wireless users and BSs, as well as the status of energy and data queues. Furthermore, for low service demands periods, DDQN can also control BSs' (de)activation to reduce overall energy consumption while maintaining QoS provisioning. A similar setting was considered for a vehicular edge computing (VEC) system [33] and DQN was employed to maximize the system's long-term utility, considering the stochastic vehicle traffic, dynamic computation requests, and time-varying channel conditions. Beyond infrastructure-based offloading, collaborative offloading enables users to offload computation tasks to nearby devices via device-to-device (D2D) communications [34]. The optimization of such collaborative policies depends on the remaining workload of each user, the availability of local computation resources, and inter-user link quality. DQN can be employed to learn an optimal offloading policy that maximizes resource utilization or minimizes total power consumption, subject to individual users' energy and delay constraints. Similarly, collaborative offloading is envisioned in a VEC system [35], allowing vehicles to collaborate with each other in computing and relaying via vehicle-to-vehicle (V2V) communications. The optimal data scheduling strategy can be derived by DQN to minimize the system-wide data processing cost.

1.2.3 DRL for UAV-Assisted Wireless Networks

Recently, UAVs have been increasingly integrated into wireless networks to enhance connectivity and service quality for terrestrial devices, emerging as a key enabler of future IoT systems [4, 5, 36, 37]. Due to hardware limitations, many low-cost IoT devices struggle to meet QoS requirements for data transmission owing to constrained energy supplies, remote deployment locations, and non-line-of-sight (NLoS) channel conditions. These challenges can be mitigated by deploying UAVs as aerial APs relay nodes to assist communications between ground users (GUs) and remote terrestrial infrastructure. The UAVs' capabilities of rapid deployment, high mobility, and three-dimensional maneuverability enable flexible and adaptive network reconfiguration over large service areas. As such, UAV-assisted wireless networks not only extend communication coverage but also enhance network capacity and resource efficiency, particularly in dense or dynamically changing IoT environments.

To fully exploit the performance gains in UAV-assisted wireless networks, UAVs' mobility must be jointly optimized with transmission control strategies, leading to profound performance improvement such as increased data rates, extended coverage area, and reduced AoI in wireless networks. By exploiting more favorable GU-to-UAV (G2U) channel conditions, often characterized by high probability of line-of-sight (LoS), UAVs can hover over areas of interest, collect data from GUs, and then forward it to remote BSs. Compared to direct GU-to-BS (G2B) links, which may suffer from NLoS propagation or severe path loss, the UAV-assisted data transmissions enables higher data rates and reduced transmit power at the GUs. By carefully planning UAVs' trajectories, network operators can improve GUs' achievable data rates and reduce end-to-end transmission latency. Furthermore, energy efficiency can be enhanced through joint optimization of GUs' transmit control and UAVs' trajectories. Beyond capacity and efficiency improvements, UAVs can also support time-sensitive applications by minimizing AoI. Strategic trajectory planning allows UAVs to periodically collect sensing data from distributed GUs in a timely manner, improving information freshness across the network, as illustrated in Fig. 1.4.

The joint optimization of UAVs' trajectories and transmission control policies constitutes a high-dimensional control problem, particularly in multi-UAV-assisted wireless networks, where strong spatiotemporal dependencies arise from correlated channel fading, user mobility, and inter-UAV interference. This challenge is further exacerbated by rapidly changing network topology, UAVs' resource limitations, and physical restrictions, as well as GUs' time-varying traffic demands and diverse QoS requirements. Each UAV's trajectory not only impacts its own energy and infor-

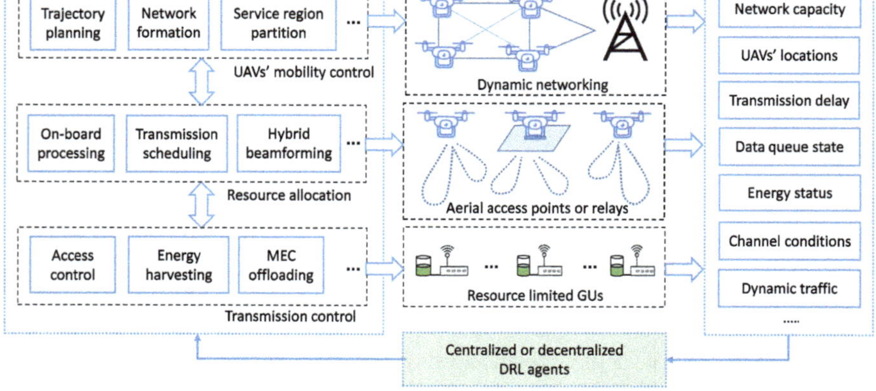

Fig. 1.4 DRL framework for UAV-assisted wireless networks

mation efficiency but also influences the current and future trajectories of other UAVs due to resource competition, interference, and spatial coordination requirements. These spatiotemporal couplings among UAVs significantly complicate the joint trajectory optimization for serving large-scale GUs. Consequently, traditional optimization methods face scalability and adaptability challenges in such complex and dynamic environment. DRL has emerged as a promising approach for optimizing UAV-assisted wireless networks [7, 20, 38, 39], capable of handling high-dimensional and coupled decision spaces. As illustrated in Fig. 1.5, DRL can be applied in multiple layers, from physical-layer resource allocation to high-level trajectory planning and networking.

- UAVs' mobility control: The mobility control mainly affects the UAVs' trajectories, leading to dynamic changes of the UAVs' connectivity. Hence, the UAVs' trajectory planning in the physical world should be closely coupled with an adaptive network formation strategy in the network domain to maintain dynamic and robust multi-hop UAV-to-UAV (U2U) relaying communications. By interacting with the network environment, DRL enables dynamic optimization of UAVs' trajectories and network formation to avoid overlapping service region and ensure more efficient data transmissions. Network formation controls the information flows among UAVs, which should be dynamically adapted based on UAVs' positions, channel conditions, energy states, and data queue sizes [40].
- On-board resource allocation: The UAVs' resource allocation aims at optimally managing the vertical information flows from GUs to the UAVs, as well as the horizontal information flows among UAVs. As the aerial APs, DRL can optimize the UAVs' transmission scheduling strategy to better serve the GUs, balancing the energy consumption, AoI performance, or throughput fairness according to diverse QoS requirements. Considering UAVs as the MEC servers, DRL can optimally allocate computational resources to assist GUs' information processing, e.g., semantic extraction and inference [41]. The UAVs can also serve as the wireless power transmitters by beamforming radio frequency (RF) energy to GUs with energy harvesting capability. DRL can optimize the UAVs' beamforming strategy to balance simultaneous wireless information and power transfer (SWIPT) for different GUs. As shown in Fig. 1.4, the UAVs can potentially carry the intelligent reflecting surface (IRS) to proactively improve channel conditions by optimizing the IRS's passive beamforming strategy [42].
- GUs' access control: DRL can be used to design efficient and adaptive channel access strategies for GUs to control their information transmissions, particularly in dynamic networks with unknown traffic demands, channel conditions, energy storage, etc. For example, depending on the energy availability and traffic demands, GUs can switch among active RF communications, low-power backscatter communications, and the energy harvesting modes [43, 44]. When GUs have sufficient energy, they may prefer active RF communications for more reliable and higher throughput. With insufficient energy supply, the GUs choose backscatter communications for spectrum efficiency or energy harvesting mode to recharge their batteries. Considering GUs as the edge users under each UAV's coverage, DRL can

help optimize GUs' semantic extraction and data offloading decisions, effectively balancing the resource demands of computation and transmission in UAV-assisted wireless networks.

A typical implementation of DRL relies on centralized control at the BS, which aggregates all UAVs' state information to jointly optimize their trajectories and transmission policies, maximizing the overall network performance. While effective, this approach incurs substantial communication and training overheads that scale with the number of UAVs. An alternative and more scalable approach assumes each UAV as an independent agent that adapts its own trajectory and transmission control based on local observations of the environment [8, 20]. This decentralized execution enables real-time adaptability with minimal communication. However, such multi-agent DRL frameworks often require centralized training, which can be communication-intensive and computationally costly in large-scale deployments.

1.3 Model-Free DRL Meets Model-Based Optimization

Conventionally, optimal network control relies on formulating performance optimization problems that explicitly characterize the relationship between network control variables and the output performance metrics. Lots of efforts in literature have been devoted to designing efficient algorithms to approximate the optimal solutions to these control problems. A fundamental limitation of this approach is the demand for an accurate modeling of the wireless network and mathematically tractable problem formulation to facilitate efficient resource allocation and implementation. Such a model-based optimization strategy nowadays faces the computational challenges and scalability issues as the future wireless networks become more diversified, decentralized, dynamic, and autonomous. Instead of the conventional optimization methods, the model-free DRL has been shown to be a flexible framework by adapting the wireless network based on continuous interactions among network entities. In the current literature, DRL's application in wireless networks is envisioned to be an flexible and robust solution that revolutionizes the network optimization paradigm [7, 8, 20, 21]. However, DRL still has obvious limitations for practical deployment due to insufficient training data or excessive overhead, especially for large-scale decentralized wireless networks.

The interplay between model-based optimization and model-free DRL can be exploited as a promising approach to enhance agile and autonomous decision making in future wireless networks [45, 46]. Model-based optimization leverages well-established prior knowledge, such as channel characteristics, signal models, access protocols, and other operational workflows, which represents empirical study, human expertise, and rule-based design principles for wireless networks. As such, model-based optimization can achieve high efficiency and even provable performance guarantees, providing theoretical performance bounds for ideal cases with complete information of the system. It also serves as reliable approximations of practical wireless networks with uncertain information. Instead, DRL operates as a data-driven or model-free method to search for optimal policies through continuous trials and errors. In particular, it randomly explores possible combinations of control variables for wireless networks and selects the best control combination based on a quality evaluation. Such a model-free DRL can be more flexibly applied to complex systems that are difficult to model explicitly and accurately. It is also more robust to fight against the uncertain dynamics of practical wireless networks.

By taking advantages of both methods, we are motivated to explore their interplay and integrated designs to facilitate resource-efficient control and deployment of future wireless networks. Specifically, to invoke further discussion, we summarize possible interplays between model-based optimization and model-free DRL as follows:

- Optimization generates initial policy: Instead of collecting training data from measurement, model-based optimization can approximate the optimal network performance given each state of the system. This helps build a high-quality and large-scale training dataset efficiently, which is then used to pre-train a policy network, providing a reasonable initial policy to accelerate subsequent DRL training. Moreover, the model-based training dataset can be integrated with the sporadic real-world measurement to fine-tune the policy for practical deployment [45].
- Optimization guides and speeds up DRL: Model-based optimization can be integrated into each decision epoch by directly providing high-quality estimates of target values or actions to improve sample efficiency and speed up convergence [46]. Given the current state, model-based optimization can be invoked to estimate the value function by relaxation, approximation, or simple heuristics. Such a model-based value estimate can replace the Q-value estimate to directly guide the training of critic network, leading to faster and more stable convergence. Similarly, model-based optimization can also generate reasonable actions efficiently. Instead of random exploration, the model-based action can narrow the search space, allowing the DRL agent to focus on high-potential exploration.
- Optimization reduces DRL's search space: We can bridge the gap between model-based optimization and model-free DRL by flexibly dividing the control variables of complex systems into two subsets. One subset is adapted by model-free DRL while the other is approximated by model-based optimization, leveraging analytical structure rather than relying solely on random exploration. By solving inner-loop optimization, the state and action spaces for the outer-loop DRL can be drastically reduced. Such a hierarchical DRL can accelerate learning performance [19, 47].
- Optimization parallelizes multi-agent DRL: Multi-agent DRL typically requires all agents' states and actions to make coordinated decisions, which drastically increases the search space and slows down convergence. Model-based optimization improves the learning efficiency by providing action estimates for individual agents as better-informed starting points for action exploration. Each agent can further refine its own action by model-free DRL, treating the behaviors of other agents as encapsulated by model-based action estimates. This approach effectively decouples the multi-agent learning process, enabling parallelized and independent

DRL updates for each agent based on local observations [48, 49]. As such, the state and action spaces can be reduced dramatically to improve the learning efficiency.

- DRL drives stochastic optimization: DRL is proficient at solving intractable optimization problems. Thus it can be integrated as an adaptive solver for stochastic optimization problems spanning over multiple time slots. Typically, stochastic optimization can be decomposed into a series of per-slot deterministic subproblems by Lyapunov optimization framework. Each subproblem can be still computational expensive by using model-based optimization. In this case, DRL can be invoked in each time slot to approximate each subproblem and thus drive the temporal evolution of such a dynamic system, ensuring asymptotic achievement of the stochastic objective and satisfaction of the long-term resource constraints [50].

The above analysis motivates our following chapters to demonstrate the interplays between model-based optimization and model-free DRL, and their integrated designs for performance optimization in various applications of future wireless networks, including MEC and UAV-assisted wireless networks.

1.4 Organization

This chapter first reviews the fundamentals of DRL and representative DRL algorithms, including DQN and its variants for discrete action spaces, as well as DDPG and PPO for continuous control in the actor-critic framework. We then briefly analyze their design principles, applicability, and computational characteristics, highlighting their potential applications and generic frameworks for various control problems in wireless networks, including decentralized access control, resource allocation in MEC systems, and UAV-assisted wireless networks. Furthermore, we focus on practical limitations of DRL and discuss potential interplays between model-based optimization and model-free DRL for accelerating learning performance. As shown in Fig. 1.5, the following chapters will demonstrate these interplays by providing specific case studies with integrated algorithm designs for various control problems in wireless networks.

In Chap. 2, we propose optimization-driven DRL methods to accelerate the model-free DRL in complex wireless networks. Leveraging partial information from various network entities, we can construct approximate optimization problems that are efficiently tractable to guide DRL in each episode. This can significantly improves learning efficiency and reward performance. The case study focuses on high dimensional beamforming optimization in an IRS-assisted wireless network, including RF transmitter's active beamforming and IRS's passive beamforming strategies. As passive beamforming is challenging to optimize directly due to its non-convex structure and lack of explicit channel models, we employ an outer-loop DDPG to search for near-optimal phase shifts. Given these phase shifts, the optimal active beamforming can be efficiently computed using conventional optimization methods. This

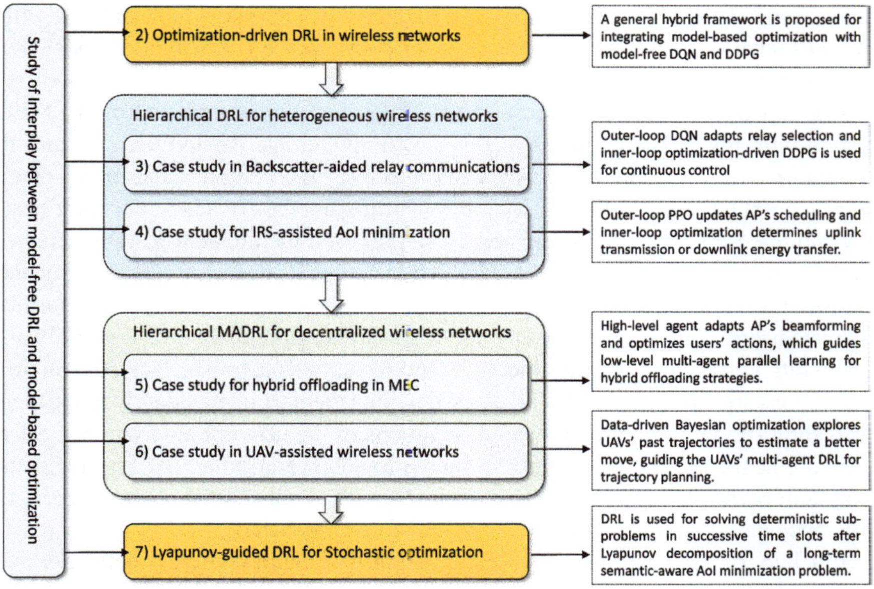

Fig. 1.5 Chapter organization, highlighting the interplay between model-free DRL and model-based optimization via case studies in wireless networks

decomposition reduces DDPG's action space by isolating the high-dimensional passive beamforming variables, thereby improving learning efficiency and convergence speed compared to model-free DDPG.

In Chap. 3, we focus on large-scale heterogeneous wireless networks and design the hierarchical DRL framework that splits the complete control variables into two parts to ensure fast convergence in decentralized decision making. One part can be iteratively updated by model-free DRL methods in the outer loop, while the other can be efficiently approximated by model-based optimization in the inner loop. The case study focuses on throughput maximization in a backscatter-aided relay communications system by a joint optimization of relay selection and transmission control. Thus, we update the combinatorial relay selection by the outer-loop DDPG. Given the relay selection, the inner-loop optimization becomes efficient to refine the beamforming and time allocation strategy. It also provides a performance lower bound that drives the DDPG agent to select a better action in each decision epoch.

In Chap. 4, we extend the hierarchical DRL framework to AoI minimization problem in an IRS-assisted wireless network. Considering dynamic channel conditions, the outer-loop PPO is employed to adapt the per-slot scheduling decision for each sensor device to upload sensing information to an AP. If no sensor device is scheduled due to energy shortage or deteriorating channel conditions, the AP will schedule downlink wireless power transfer to all sensor devices with energy harvesting capabilities. Given the outer-loop scheduling decision generated by PPO, the inner-loop optimization refines the transmission control parameters by simple relaxation and approximations. Such a hierarchical PPO is shown to achieve significantly higher cumulative rewards and faster convergence compared to model-free DRL approaches.

In Chap. 5, the case study focuses on MADRL for coordinated control in a MEC system with multiple edge users of diverse computational capabilities. We aim to maximize overall energy efficiency while ensuring the complete execution of all users' computational workloads, by jointly optimizing the AP's beamforming strategy and the users' offloading decisions. The proposed hierarchical MADRL framework includes the high-level agent at the AP to update its beamforming strategy and the low-level user agents to adapt their offloading strategies. The AP can further estimate users' actions by efficiently solving an approximate optimization problem, leveraging knowledge of their objectives and resource constraints. The AP's action estimations are then broadcast to all users, enabling them to independently refine their offloading strategies based on local observations of the network environment. Such a hierarchical MADRL framework is shown to achieve up to a 50% improvement in system performance over baseline methods, with significantly enhanced learning efficiency and reliability compared to model-free approaches.

In Chap. 6, we tackle uncertain information in dynamic wireless networks due to time-varying channel conditions, mobility, or measurement delays, which motivates the Bayesian-enhanced DRL for more robust learning by leveraging Bayesian inference. The case study focuses on a UAV-assisted wireless network to help relay GUs' sensing information to remote BS. We use MADRL to adapt UAVs' trajectory planning and network formation strategies that minimize overall energy consumption and end-to-end transmission delay. Considering limited information exchange among UAVs, we further leverage Bayesian optimization for each UAV to infer a more preferable hovering location for data collection based on its historical trajectory observations. This reduces inefficient action exploration during the early stages of training and significantly improves convergence speed in the MADRL framework.

In Chap. 7, we address long-term time-averaged performance optimization in dynamic wireless networks, which is usually formulated as a stochastic optimization problem with high computational complexity. We introduce a Lyapunov-guided DRL framework that leverages the Lyapunov optimization method to decompose the multi-stage stochastic problem into a sequence of per-slot subproblems. Each subproblem is then solved adaptively using DRL based on real-time network state observations. The case study focuses on a UAV-assisted wireless network, where GUs report sensing data to a remote BS via UAVs' relaying. Both GUs and UAVs are equipped with semantic extraction capabilities, enabling them to distill meaningful information from raw sensing data and thus improve transmission efficiency. We formulate a stochastic semantic-aware AoI minimization problem by jointly optimizing UAV-GU associations, semantic information extraction, and UAVs' trajectories under mobility and communication constraints.

References

1. Y. Mao, C. You, J. Zhang, K. Huang, K.B. Letaief, A survey on mobile edge computing: the communication perspective. IEEE Commun. Surv. Tutor. **19**(4), 2322–2358 (2017)
2. L. Lei, Y. Tan, K. Zheng, S. Liu, K. Zhang, X. Shen, Deep reinforcement learning for autonomous internet of things: model, applications and challenges. IEEE Commun. Surv. Tutor. **22**(3), 1722–1760 (2020)
3. Z. Zhou, X. Chen, E. Li, L. Zeng, K. Luo, J. Zhang, Edge intelligence: paving the last mile of artificial intelligence with edge computing. Proc. IEEE **107**(8), 1738–1762 (2019)
4. Y. Zeng, Q. Wu, R. Zhang, Accessing from the sky: a tutorial on UAV communications for 5G and beyond. Proc. IEEE **107**(12), 2327–2375 (2019)
5. A. Fotouhi, H. Qiang, M. Ding, M. Hassan, L.G. Giordano, A. Garcia-Rodriguez, J. Yuan, Survey on UAV cellular communications: practical aspects, standardization advancements, regulation, and security challenges. IEEE Commun. Surv. Tutor. **21**(4), 3417–3442 (2019)
6. K.B. Letaief, W. Chen, Y. Shi, J. Zhang, Y.-J.A. Zhang, The roadmap to 6G: AI empowered wireless networks. IEEE Commun. Mag. **57**(8), 84–90 (2019)
7. N.C. Luong, D.T. Hoang, S. Gong, D. Niyato, P. Wang, Y.-C. Liang, D.I. Kim, Applications of deep reinforcement learning in communications and networking: a survey. IEEE Commun. Surv. Tutor. **21**(4), 3133–3174 (2019)
8. A. Feriani, E. Hossain, Single and multi-agent deep reinforcement learning for AI-enabled wireless networks: a tutorial. IEEE Commun. Surv. Tutor. **23**(2), 1226–1252 (2021)
9. R.S. Sutton, A.G. Barto, *Reinforcement Learning: An Introduction* (The MIT Press, Cambridge, MA, 2018)
10. S. Gong, Y. Xie, J. Xu, D. Niyato, Y.-C. Liang, Deep reinforcement learning for backscatter-aided data offloading in mobile edge computing. IEEE Network **34**(5), 106–113 (2020)
11. V. Mnih, K. Kavukcuoglu, D. Silver, A.A. Rusu, J. Veness, M.G. Bellemare, A. Graves, M. Riedmiller, A.K. Fidjeland, G. Ostrovski, S. Petersen, C. Beattie, A. Sadik, I. Antonoglou, H. King, D. Kumaran, D. Wierstra, S. Legg, D. Hassabis, Human-level control through deep reinforcement learning. Nature **518**(7540), 529–533 (2015)
12. T.P. Lillicrap, J.J. Hunt, A. Pritzel, N. Heess, T. Erez, Y. Tassa, D. Silver, D. Wierstra, Continuous control with deep reinforcement learning, in *Proceedings of the International Conference Learning Representations (ICLR)*, San Juan, Puerto Rico, May 2016
13. H. Van Hasselt, A. Guez, D. Silver, Deep reinforcement learning with double Q-learning, in *Proceedings of the AAAI Conference Artificial Intelligence*, Phoenix, Arizona, USA, Feb 2016, pp. 2094–2100
14. Z. Wang, T. Schaul, M. Hessel, H. Van Hasselt, M. Lanctot, N. De Freitas, Dueling network architectures for deep reinforcement learning, in *Proceedings of the International Conference on Machine Learning (ICML)*, New York, NY, USA, Jun 2016, pp. 1995–2003
15. T. Schaul, J. Quan, I. Antonoglou, D. Silver, Prioritized experience replay, in *Proceedings of the International Conference Learning Representations*, Puerto Rico, May 2016
16. D. Kangin, N. Pugeault, On-policy trust region policy optimisation with replay buffers, in *Proceedings of International Conference on Learning Representations (ICLR)*, New Orleans, LA, USA, May 2019
17. J. Schulman, S. Levine, P. Abbeel, M. Jordan, P. Moritz, Trust region policy optimization, in *Proceedings of the International Conference Machine Learning (ICML)*, Jul 2015, pp. 1889–1897
18. J. Schulman, F. Wolski, P. Dhariwal, A. Radford, O. Klimov, Proximal policy optimization algorithms. arXiv:1707.06347 (2017)
19. S. Gong, L. Cui, B. Gu, B. Lyu, D.T. Hoang, D. Niyato, Hierarchical deep reinforcement learning for age-of-information minimization in IRS-aided and wireless-powered wireless networks. IEEE Trans. Wirel. Commun. **22**(11), 8114–8127 (2023)
20. T. Li, K. Zhu, N.C. Luong, D. Niyato, Q. Wu, Y. Zhang, B. Chen, Applications of multi-agent reinforcement learning in future internet: a comprehensive survey. IEEE Commun. Surv. Tutor. **24**(2), 1240–1279 (2022)

21. W. Chen, X. Qiu, T. Cai, H.-N. Dai, Z. Zheng, Y. Zhang, Deep reinforcement learning for internet of things: a comprehensive survey. IEEE Commun. Surv. Tutor. **23**(3), 1659–1692 (2021)
22. Y. Yu, T. Wang, S.C. Liew, Deep-reinforcement learning multiple access for heterogeneous wireless networks. IEEE J. Selected Area. Commun. **37**(6), 1277–1290 (2019)
23. J. Park, S. Samarakoon, M. Bennis, M. Debbah, Wireless network intelligence at the edge. Proc. IEEE **107**(11), 2204–2239 (2019)
24. W. Xu, Z. Yang, D.W.K. Ng, M. Levorato, Y.C. Eldar, M. Debbah, Edge learning for B5G networks with distributed signal processing: Semantic communication, edge computing, and wireless sensing. IEEE J. Selected Topic. Signal Process. **17**(1), 9–39 (2023)
25. H. Zhu, Y. Cao, W. Wang, T. Jiang, S. Jin, Deep reinforcement learning for mobile edge caching: review, new features, and open issues. IEEE Netw. **32**(6), 50–57 (2018)
26. N. Abbas, Y. Zhang, A. Taherkordi, T. Skeie, Mobile edge computing: a survey. IEEE Internet Things J. **5**(1), 450–465 (2018)
27. X. Wang, Y. Han, V.C.M. Leung, D. Niyato, X. Yan, X. Chen, Convergence of edge computing and deep learning: a comprehensive survey. IEEE Commun. Surv. Tutor. **22**(2), 869–904 (2020)
28. C. Zhang, Z. Liu, B. Gu, K. Yamori, and Y. Tanaka, A deep reinforcement learning based approach for cost-and energy-aware multi-flow mobile data offloading. IEICE Trans. Commun. **E101-B**(7), 2017–2025 (2018)
29. M. Min, L. Xiao, Y. Chen, P. Cheng, D. Wu, W. Zhuang, Learning-based computation offloading for IoT devices with energy harvesting. IEEE Trans. Veh. Technol. **68**(2), 1930–1941 (2019)
30. L. Huang, S. Bi, Y.-J.A. Zhang, Deep reinforcement learning for online computation offloading in wireless powered mobile-edge computing networks. IEEE Trans. Mobile Comput. **19**(11), 2581–2593 (2020)
31. X. Chen, H. Zhang, C. Wu, S. Mao, Y. Ji, M. Bennis, Optimized computation offloading performance in virtual edge computing systems via deep reinforcement learning. IEEE Internet Things J. **6**(3), 4005–4018 (2019)
32. M. Tang, V.W. Wong, Deep reinforcement learning for task offloading in mobile edge computing systems. IEEE Trans. Mobile Comput. **21**(6), 1985–1997 (2022)
33. Y. Liu, H. Yu, S. Xie, Y. Zhang, Deep reinforcement learning for offloading and resource allocation in vehicle edge computing and networks. IEEE Trans. Veh. Technol. **68**(11), 11158–11168 (2019)
34. D.V. Le, C.-K. Tham, A deep reinforcement learning based offloading scheme in ad-hoc mobile clouds, in *Proceedings of IEEE INFOCOM (Workshop)*, Honolulu, USA, Apr 2018
35. Q. Luo, C. Li, T.H. Luan, W. Shi, Collaborative data scheduling for vehicular edge computing via deep reinforcement learning. IEEE Internet Things J. **7**(10), 9637–9650 (2020)
36. M. Erdelj, E. Natalizio, K.R. Chowdhury, I.F. Akyildiz, Help from the sky: leveraging UAVs for disaster management. IEEE Pervasive Comput. **16**(1), 24–32 (2017)
37. M. M. Azari, S. Solanki, S. Chatzinotas, O. Kodheli, H. Sallouha, A. Colpaert, J. F. Mendoza Montoya, S. Pollin, A. Haqiqatnejad, A. Mostaani, E. Lagunas, B. Ottersten, Evolution of non-terrestrial networks from 5G to 6G: a survey. IEEE Commun. Surv. Tutor. **24**(4), 2633–2672 (2022)
38. X. Liu, Y. Liu, Y. Chen, Reinforcement learning in multiple-UAV networks: deployment and movement design. IEEE Trans. Veh. Technol. **68**(8), 8036–8049 (2019)
39. N. Zhao, Z. Ye, Y. Pei, Y.-C. Liang, D. Niyato, Multi-agent deep reinforcement learning for task offloading in UAV-assisted mobile edge computing. IEEE Trans. Wirel. Commun. **21**(9), 6949–6960 (2022)
40. S. Zhang, H. Zhang, B. Di, L. Song, Cellular UAV-to-X communications: design and optimization for multi-UAV networks. IEEE Trans. Wirel. Commun. **18**(2), 1346–1359 (2019)
41. N. Cheng, F. Lyu, W. Quan, C. Zhou, H. He, W. Shi, X. Shen, Space/aerial-assisted computing offloading for IoT applications: a learning-based approach. IEEE J. Selected Area Commun. **37**(5), 1117–1129 (2019)
42. S. Zhao, S. Gong, B. Gu, L. Li, B. Lyu, D. Thai Hoang, C. Yi, Exploiting NOMA transmissions in multi-UAV-assisted wireless networks: from aerial-RIS to mode-switching UAVs. IEEE Trans. Wirel. Commun. **24**(3), 2530–2544 (2025)

43. S. Gong, L. Gao, J. Xu, Y. Guo, D.T. Hoang, D. Niyato, Exploiting backscatter-aided relay communications with hybrid access model in device-to-device networks. IEEE Trans. Cogn. Commun. Netw. **5**(4), 835–848 (2019)
44. C. Chen, S. Gong, B. Gu, L. Li, Z. Liu, N. C. Luong, D. Niyato, Learning RIS-assisted hybrid NOMA transmissions for capacity maximization in wireless-powered networks. IEEE Trans. Veh. Technol. **75**(1), 1492–1506 (2026)
45. A. Zappone, M. Di Renzo, M. Debbah, T.T. Lam, X. Qian, Model-aided wireless artificial intelligence: embedding expert knowledge in deep neural networks for wireless system optimization. IEEE Veh. Technol. Mag. **14**(3), 60–69 (2019)
46. S. Gong, J. Lin, B. Ding, D. Niyato, D.I. Kim, M. Guizani, When optimization meets machine learning: the case of IRS-assisted wireless networks. IEEE Netw. **36**(2), 190–198 (2022)
47. S. Gong, Y. Zou, J. Xu, D.T. Hoang, B. Lyu, D. Niyato, Optimization-driven hierarchical learning framework for wireless powered backscatter-aided relay communications. IEEE Trans. Wirel. Commun. **21**(2), 1378–1391 (2022)
48. H. Zhou, Y. Long, S. Gong, K. Zhu, D.T. Hoang, D. Niyato, Hierarchical multi-agent deep reinforcement learning for energy-efficient hybrid computation offloading. IEEE Trans. Veh. Technol. **72**(1), 986–1001 (2023)
49. S. Gong, M. Wang, B. Gu, W. Zhang, D.T. Hoang, D. Niyato, Bayesian optimization enhanced deep reinforcement learning for trajectory planning and network formation in multi-UAV networks. IEEE Trans. Veh. Technol. **72**(8), 10933–10948 (2023)
50. Y. Long, S. Gong, S. Sun, G.C.F. Lee, L. Li, D. Niyato, Lyapunov-guided deep reinforcement learning for semantic-aware AoI minimization in UAV-assisted wireless networks. IEEE Trans. Wirel. Commun. **24**(8), 6351–6364 (2025)

Chapter 2
Optimization-Driven DRL in Wireless Networks

2.1 Introduction

Traditionally, efficient resource management in wireless networks relies on precise modeling and tractable problem formulations to enable centralized and model-based optimization. These methods often employ simplifications, approximations, or heuristics design to ensure computational tractability. As future wireless networks become more heterogenous, decentralized, and autonomous, such model-based optimization approaches face growing limitations in scalability and adaptability. DRL has attracted considerable attention due to superior adaptability and resilience for network optimization with uncertain information, incomplete modeling, and time-varying conditions. However, it is still a challenging task to practically deploy DRL algorithms in wireless networks. Typically, the offline training often demands extensive data and computational resources, while online learning suffers from slow convergence, especially in large-scale wireless networks with huge state and action spaces. Though model-free DRL methods can learn and update policy by continuously interacting with the network environment, they often suffer from slow convergence and inefficient exploration in complex network scenarios.

These practical challenges motivate us to design more efficient and scalable DRL approaches for future autonomous wireless systems by integrating the network or model information into the model-free DRL framework. Incorporating empirical knowledge or partial information of the system models can significantly improve learning efficiency and decision-making agility. Such information can help reduce state and action spaces, avoiding inefficient action exploration. In the case study of this chapter, we propose an optimization-driven DRL framework that leverages conventional model-based optimization to guide and accelerate DRL training. Given partial network information, e.g., the beamforming or power control strategy of the base station, model-based optimization is used to fast adapt the other control decisions, e.g., the wireless users' channel access and resource allocation strategies. Instead of random trials and errors, the model-based optimization can quickly

S. Gong et al., *Optimization-Driven Deep Reinforcement Learning for Wireless Networks*, Wireless Networks, https://doi.org/10.1007/978-3-032-22997-7_2

determine the most plausible control decisions by explicitly formulating the dependencies between the control decisions into the model-based optimization problem. As an illustrative example, in this chapter we apply the proposed framework for network performance optimization in a complex wireless system involving an IRS with high-dimensional control variables. This case study demonstrates the potential of the proposed DRL framework to enhance learning efficiency and foster practical deployment in real-world wireless networks.

The integrated design of model-based optimization and model-free DRL has received limited attention in the literature. Conventional optimization methods rely on explicit problem formulations by exploiting the inherent structural properties of the network control problems, such as deterministic relationships and physical coupling among control variables. These structural properties offer rich information to improve learning efficiency and network performance. The key idea of optimization-driven DRL is to take both advantages of model-based optimization and model-free DRL. In particular, we leverage the precision and efficiency of model-based optimization methods while preserving the adaptability and robustness of model-free DRL under dynamic and uncertain environments. The concept of model-aided AI has been previously discussed in wireless networks [1]. Model-based optimization can be simply used to create large data-sets for DNN training in the offline phase, e.g., [2] and [3]. In contrast to traditional model-aided approaches, the proposed optimization-driven DRL framework integrates model-based optimization directly into the online learning phase, dynamically guiding the learning process toward more favorable policies and improved reward performance [4]. This hybrid paradigm combines the adaptability of model-free learning with the efficiency of analytical model-based optimization.

Specifically, we employ a model-free DRL in the outer loop to ensure robustness against uncertain system dynamics, incomplete state information, and modeling inaccuracies. Simultaneously, in the inner loop, we incorporate lightweight model-based optimization methods during online training to adapt a part of the control variables, such as beamforming, reflection coefficients, and time allocation, by efficiently solving simplified or approximate versions of the underlying optimization problem. This enables faster convergence and more accurate action selection without sacrificing adaptability. Compared to prior works on supervised [2], unsupervised [3], or offline model-aided DRL methods [1], our framework is particularly suited for complex wireless networks with both inaccurate system modeling and intractable solutions. Unlike conventional DRL often suffering from slow convergence, sample inefficiency, and training instability, optimization-driven DRL accelerates learning by embedding domain knowledge, such as analytical performance bounds or partial model information, into the model-free learning loop. Moreover, in contrast to offline model-aided strategies that rely on pre-training with large synthetic datasets, the proposed optimization-driven DRL framework enables continuous online refinement, allowing the agent to adapt in real time to the changing network conditions. This makes it a powerful tool for dynamic and resource-constrained wireless networks requiring both high performance and rapid responsiveness.

In the following sections, we first analyze the limitations of conventional DRL approaches in the context of wireless network optimization, emphasizing their challenges with slow convergence, sample inefficiency, and instability in high-dimensional and dynamic network environments. Then, we present the optimization-driven DRL framework in detail, discussing its design principles, hierarchical architecture, and integration of model-based optimization into the online learning process. The algorithm design and case study were originally introduced in our earlier work [4]. This chapter refines and reorganizes the content from [4] to ensure consistency, clarity, and coherence within the broader scope of the book.

2.2 Limitations of Model-Free DQN and DDPG

DQN and DDPG are two popularly applied DRL algorithms that employ DNNs to represent the parameterized value and policy functions, respectively. To differentiate the traditional DRL algorithms from our designs, we refer to the classic DQN and DDPG as two model-free DRL approaches, which do not require explicit model information about the network environment. In wireless networks, model-free DQN has been applied to various discrete decision-making problems such as channel allocation, transmit scheduling, relay selection, and user association. The success of DQN relies on two important techniques: experience replay and fixed target Q-network, as discussed in Chap. 1. Experience replay emphasizes an efficient utilization of historical action-state transition samples by storing them in a memory buffer and randomly sampling mini-batches from the reusable memory buffer for DNN training in each decision epoch. This mechanism breaks the correlation among sequential samples and improves convergence and reliability. The target Q-network is a delayed copy of the online Q-network and used to generate more stable target Q-values during training. The DNN parameters of the target Q-network are periodically updated from the online Q-network after every a few training steps. Figure 2.1a illustrates the typical framework of model-free DQN. DDPG extends its applicability to continuous action spaces, making it suitable for more complex control tasks such as adjusting RF transmit power or tuning the IRS's phase shifts. DDPG also employs a separate target Q-network to stabilize training. Besides the Q-network, DDPG provides another set of DNNs to approximate the parameterized policy function directly. DNN approximations for the Q-value and the policy function are also named as the critic and actor networks, respectively. Figure 2.1b illustrates the typical framework of model-free DDPG, where both critic and actor networks maintain their target networks, whose parameters are slowly updated from their online versions.

The target Q-networks in both DQN and DDPG are used to stabilize the learning curves. However, the strong dependency between them may also lead to a slow learning efficiency. First, in the early stages of learning, both the online and target Q-networks are randomly initialized and typically far from their optimal values. This can lead to inaccurate Q-value estimates, which may misguide policy updates and destabilize the learning process. As a result, model-free DQN and DDPG often require a large number of historical transitions and extensive action exploration to converge toward meaningful policies, leading to a prolonged warm-up period and

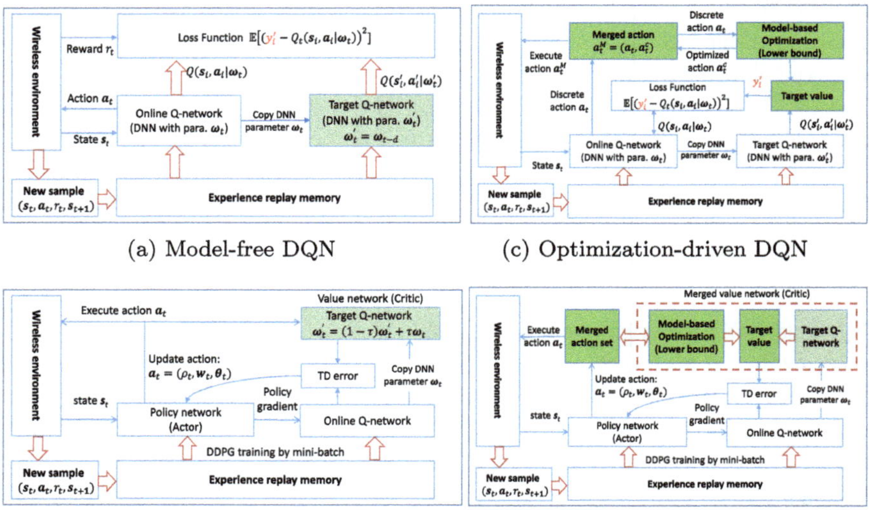

(a) Model-free DQN (c) Optimization-driven DQN

(b) Model-free DDPG (d) Optimization-driven DDPG

Fig. 2.1 Model-free vs. optimization-driven DQN/DDPG framework [4]. The optimization-driven DRL framework integrates model-based optimization into model-free DRL by decomposing control variables. One subset is optimized via DRL, the other through numerical optimization methods. This reduces action space dimensionality, enhances sample efficiency, and stabilizes learning by providing informed, decoupled targets, effectively balancing adaptability and computational efficiency for complex control problems in wireless networks

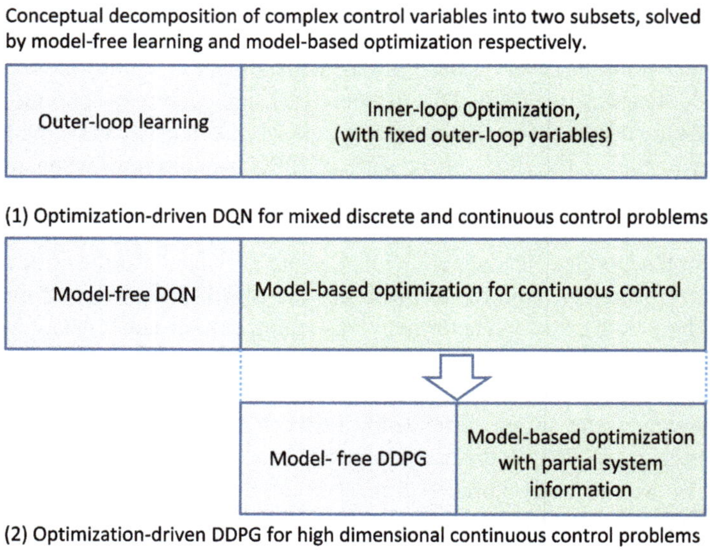

Fig. 2.2 Conceptual decomposition of control variables into two subsets, solved by model-free learning and model-based optimization respectively

low sample efficiency. Second, the delayed update of the target Q-network introduces additional challenges. For instance, DDPG's target network is updated via a soft updating rule using a fixed averaging coefficient (commonly denoted as τ), as shown in Fig. 2.1b. While this updating rule enhances training stability, the performance is highly sensitive to the choice of τ. When τ is set to a small value, the target Q-network changes slowly, which enhances stability but can significantly slow down the learning progress by delaying the incorporation of learned knowledge. On the other hand, a large τ may lead to unstable learning due to rapid propagation of estimation errors, causing performance fluctuations and even instability in training. While DQN and DDPG have demonstrated strong capabilities for various network optimization problems, this trade-off makes hyperparameter tuning difficult and practically limits their learning performance, as it often requires extensive trial-and-error especially in complex and dynamic network environments. These limitations highlight the need for more intelligent initialization and guidance for action exploration, motivating the development of optimization-driven DRL, where model-based optimization is integrated to provide informed priors to accelerate convergence and improve training stability.

2.3 Optimization-Driven DQN and DDPG

In this part, the optimization-driven DRL is proposed to enhance learning efficiency by integrating model-based optimization into a model-free DRL framework. The primary goal is to stabilize and accelerate learning by evaluating the target Q-value in a more informed and decoupled manner. This design is moativated by the following observations. Firstly, in high-dimensional and decentralized wireless networks, it is often impractical to formulate a comprehensive optimization problem that captures the full interdependencies among all control variables. Even when such a formulation is feasible, the resulting problem is typically high-dimensional and computationally intractable using conventional optimization techniques, or its solution relies on idealized assumptions, such as perfect and complete network knowledge. However, when a subset of control variables is fixed, the remaining variables can be efficiently optimized in an approximate subproblem by leveraging the inherent physical coupling among control variables.

The core design principle is to decompose the control variables into two distinct subsets, as illustrated in Fig. 2.2. One subset is determined in the outer loop using model-free DQN or DDPG, operating over a reduced action space, while the other subset is optimized instantaneously and efficiently in an inner loop, conditioned on the outer-loop control variables. Consequently, the way for decomposing the control variables becomes a critical design choice. It directly influences both the learning performance of the outer-loop DRL and the computational complexity of the inner-loop optimization. A key intuition behind this decomposition is to decouple intertwined control dependencies, ensuring that the inner-loop problem remains tractable, lightweight, and amenable. To enhance robustness, especially in the presence of modeling inaccuracies such as imperfect channel information, the inner-loop optimization should exhibit low sensitivity to variations in system parameters. In the following, we present two typical implementations of optimization-driven DRL for complex high-dimensional control problems in wireless networks.

Optimization-driven DQN: For mixed decision-making problems involving both discrete and continuous variables, we decompose the action vector into two components: the discrete actions $\mathbf{a}_t$ and the continuous actions $\mathbf{a}_t^c$, as shown in Fig. 2.1c. The discrete actions are determined by model-free DQN in the outer loop, while the continuous actions are optimized in the inner loop using numerical optimization techniques. Specifically, given the selected discrete actions, the continuous subproblem is approximated as a convex optimization problem, which can be solved efficiently. The resulting continuous actions are then combined with the discrete ones to form the complete action $(\mathbf{a}_t, \mathbf{a}_t^c)$ executed in the wireless environment. A representative application of this framework is relay selection in an IRS-assisted wireless network. The outer-loop DQN is deployed to select the optimal relay node (discrete decision), while the inner-loop optimization computes the joint active beamforming at the transmitter and the IRS's passive beamforming (continuous decisions), given the outer-loop relay selection.

Optimization-driven DDPG: For high-dimensional continuous control problems common in wireless networks, simultaneously optimizing all control variables is often infeasible due to the complexity and non-convex problem structure. Model-free DRL also struggles in such settings as large continuous action and state spaces lead to slow convergence and poor sample efficiency. To address these challenges, we extend the divide-and-conquer principle of the optimization-driven framework to continuous control problems by combining model-free DRL with model-based optimization. Specifically, we decompose the action vector into two subsets. One subset is learned by model-free DDPG in the outer loop, while the other is computed in an inner-loop optimization module. As illustrated in Fig. 2.1d, once the outer-loop DDPG outputs a partial action, the optimization module computes the remaining actions by solving a simplified, often convexified approximation of the original problem. This not only reduces the dimensionality but also provides a performance lower bound to the original control problem. The key to effective decomposition lies in the design of action splitting strategy. It should exploit structural couplings among control variables so that the inner-loop optimization remains computationally lightweight and robust. Ideally, fixing part of the action should render the remaining subproblem efficiently solvable using standard optimization tools. A practical instance of this framework arises in joint active and passive beamforming for IRS-assisted wireless networks. The joint optimization of active and passive beamforming results in a highly non-convex and coupled problem. However, when the IRS's passive beamforming is fixed, the RF transmitter's active beamforming can be well approximated as a convex problem. This motivates a decomposition where the outer-loop DDPG learns the passive beamforming strategy, while the inner-loop optimization efficiently computes the active beamforming solution. This action splitting significantly accelerates learning while maintaining high policy quality.

The key features of the proposed optimization-driven DRL framework are summarized as follows to highlight its benefits in wireless control problems.

- The optimization-driven DRL framework provides a general and flexible approach for solving complex network control problems characterized by uncertain system dynamics, imperfect modeling, and high computational complexity. By integrating a model-based optimization module into the model-free learning process, the framework guides the policy search toward higher rewards in a more stable and sample-efficient manner. The effectiveness of the framework hinges on the strategic decomposition of control variables that determines both the tractability of the inner-loop optimization and the reduced action space of the outer-loop learning agent. Compared to model-free DRL, the optimization-driven DRL significantly reduces the dimensionality of the decision space, alleviates the burden of action exploration, and enhances learning speed and policy performance.
- The partitioning of control variables in the optimization-driven DRL framework can be tailored according to practical constraints such as computational resources, training overhead, and desired solution accuracy. A principled approach is to assign variables with strong, well-defined physical or mathematical dependencies to the inner-loop optimization module. In contrast, variables with implicit, nonlinear, or poorly understood dependencies can be left to the outer-loop DRL agent, which learns effective policies through interaction and exploration. In one extreme, all control variables are handled by the DRL agent, reducing the system to conventional model-free DRL, which is flexible but often slow to converge and computationally intensive in high-dimensional spaces. In the opposite extreme, the entire control problem is solved via model-based optimization, which may yield optimal solutions under ideal conditions but lacks adaptability in dynamic and uncertain environments. It also becomes computationally prohibitive for large-scale non-convex problems. Thus, the optimization-driven DRL framework strikes a flexible balance between these two extreme cases. It leverages the efficiency and structure-awareness of model-based optimization while preserving the robustness and adaptability of model-free DRL, making it scalable, stable, and efficient for complex control problems in wireless networks.
- We introduce a novel online integration of model-based optimization and model-free DRL. Existing works often employ model-based optimization to generate synthetic datasets for offline training [1], treating optimization as a pre-processing or data generation tool. In contrast, our framework embeds the optimization module directly into the online decision-making loop, allowing it to dynamically inform and guide the learning process in real time. Unlike supervised learning, where target outputs are explicitly provided by the environment, the optimization-driven target in our framework is implicitly estimated by the DRL agent using local observations during online learning. Given the partially observed system state, the inner-loop optimization approximates the original problem and determines a feasible subset of control actions. This solution provides a performance lower bound on the optimal policy and serves as a more informed and physically meaningful estimate for the target Q-value. Particularly in the early stages of training, when

the Q-network is poorly initialized, this optimization-driven target offers a reliable learning signal that accelerates convergence. Furthermore, the optimization-driven target is decoupled from both the online and target Q-networks in model-free DRL. This independence enhances stability, as the optimization-driven target is not subject to the same parameter fluctuations of the online Q-network. As such, our approach mitigates performance oscillations, achieving robust and faster stabilization of the learning process.

In the following, we apply the proposed optimization-driven DRL to maximize network performance in a complex IRS-assisted wireless network characterized by high-dimensional and coupled control variables. Through a detailed case study, we demonstrate that the proposed approach achieves significantly higher rewards while requiring substantially fewer learning episodes compared to conventional model-free DRL. This improvement highlights the superior sample efficiency, faster convergence, and enhanced learning stability of optimization-driven DRL in resource-constrained wireless networks.

2.4 Case Study in IRS-Assisted Wireless Networks

Traditionally, the radio environment is treated as an uncontrollable and exogenous factor, and thus is often excluded from explicit consideration in network optimization formulations. Due to its inherent randomness, wireless signal propagation is subject to various physical phenomena, including reflections, diffractions, scattering, and multipath fading, before reaching the receiver. These effects collectively induce random channel fading, which significantly degrades signal quality and becomes a fundamental bottleneck in the performance optimization of wireless networks. Recently, IRS has been introduced in wireless networks to enhance information transmission performance by actively configuring the signal propagation environment [5]. The IRS's passive beamforming, i.e., achieved through a joint phase control of its scattering elements, can be exploited to actively manipulate the reflection of incident signals from different paths. By intelligently aligning multipath components, the IRS can constructively enhance the desired signal or destructively suppress interference at the receiver. Consequently, the optimization of the IRS's passive beamforming enables the creation of more favorable channel conditions, leading to significant improvements in network performance, such as increased channel capacity and reduced transmit power consumption. IRS-assisted wireless networks thus fundamentally expand the controllability of the wireless propagation environment. However, this new degree of freedom also introduces complex and non-convex optimization challenges, particularly in large-scale and dynamic networks, therefore demanding novel and efficient approaches for network performance optimization.

The optimization of IRS's passive beamforming faces significant practical challenges, particularly as the number of scattering elements increases. With a large size of IRS, the dimensionality of the phase shifting control space grows rapidly, leading to prohibitively high computational complexity that makes it difficult for real-time and scalable implementation. Moreover, conventional optimization approaches rely heavily on accurate channel information, including both transmitter-IRS and IRS-receiver links. However, lacking dedicated sensing or signal processing capabilities, the channel estimation for passive IRS is inherently challenging and often results in imperfect or outdated channel estimates. The practical limitations for channel information estimation often result in channel uncertainty and correspondingly unreliable or conservative optimization solutions. Therefore, performance optimization in IRS-assisted wireless networks demands a more careful balance between model accuracy and tractability. This often involves model simplifications, approximations, and the development of sophisticated or heuristic algorithms capable of handling non-convex, large-scale problems under imperfect information. This motivates the integration of data-driven approaches and learning techniques with classical optimization theory to efficiently cope with the complexity and dynamics of future wireless environments.

2.4.1 An Overview of IRS-Assisted Wireless Networks

By integrating the intelligent radio environment into network optimization, IRS-assisted wireless networks are expected to revolutionize the traditional network design paradigm and play an pivotal role in future wireless networks [6, 7]. Inspired by a comprehensive literature survey in [5], we can briefly summarize the key benefits of IRS-assisted wireless networks as follows:

- Easy deployment and sustainable operations: The IRS consists of low-cost, passive scattering elements integrated into a metasurface, enabling simple and non-invasive installation on building facades, indoor walls, ceilings, and other urban infrastructure. This facilitates flexible deployment and reconfiguration without major structural modifications. Moreover, IRS does not require active components for signal amplification or processing and thus it consumes negligible power and can operate in a battery-free manner by leveraging ambient RF energy harvesting. This supports sustainable and long-term operation in energy-constrained wireless networks.
- Flexible reconfiguration via passive beamforming: By jointly optimizing the phase shifts of numerous passive scattering elements, an IRS can coherently steer and focus reflected signals toward desired receivers, effectively shaping the wireless propagation environment. Its deployment offers substantial potential for enhancing key performance metrics such as signal strength, coverage, and spectral efficiency. Furthermore, the IRS's phase configuration can be jointly optimized with RF transceiver's control parameters, including active beamforming, power allocation, and scheduling, enabling a holistic optimization framework to fully exploit the complementary operations of active and passive radios in future wireless networks.
- Enhanced capacity and network performance: By leveraging passive beamforming, IRS-assisted networks can reconfigure the wireless channel to achieve higher

data rates while reducing transmit power, thereby improving energy and spectral efficiency. The IRS can also enhance interference management by destructively combining interfering signals at unintended receivers. As a result, the IRS-assisted wireless networks enable superior QoS provisioning and significantly improve key performance metrics such as sum-rate throughput and max-min fairness across wireless users.

- Enabling emerging wireless applications: The IRS is promising to unlock novel research frontiers and emerging applications in wireless networks. For instance, by jointly optimizing active and passive beamforming, the IRS can enhance physical-layer security by intentionally degrading the channel quality at potential eavesdroppers, while strengthening the legitimate signal. Beyond secure communications, IRS holds significant promise for emerging paradigms such as SWIPT, MEC offloading with radio mode switching, and UAV-assisted wireless networks, where the IRS can provide reliable, reconfigurable, and energy-efficient links in dynamic environments.

Conventionally, realizing the full performance potential of IRS-assisted wireless networks, such as enhanced channel capacity, reduced transmit power, or improved secrecy, requires a joint optimization of active beamforming at the RF transceiver and passive beamforming at the IRS. However, such optimization approaches generally face several practical challenges as follows [4]:

- Computational complexity: As the number of IRS's scattering elements increases, the dimensionality of the joint active and passive beamforming optimization grows significantly. This leads to high computational complexity, making real-time optimization challenging. Consequently, sophisticated algorithmic designs involving iterative approximations or problem decomposition are often required, which may still fall short of meeting resource constraints in dynamic networks.
- Dynamic channel conditions: Accurate channel information is essential for effective beamforming optimization. However, in large-scale IRS's deployments, the overhead for channel estimation becomes prohibitively high due to the massive number of passive elements. Moreover, in fast-varying environments, channel information is often outdated or imperfect, leading to suboptimal beamforming configurations. These estimation errors degrade system performance and destabilize model-based optimization.
- Imprecise modeling: The joint beamforming optimization typically relies on simplified and idealized system models to maintain tractability, such as static channels, continuous phase shifts, perfect hardware implementation, or convex problem formulations. However, these assumptions hardly hold in practical systems. As a result, optimization solutions derived from such models only approximate the ordinal control problems, limiting their practical effectiveness and robustness in real-world deployments.

These limitations of conventional optimization methods, including high computational complexity, sensitivity to channel dynamics, and reliance on idealized models, motivate the exploration of model-free ML approaches for solving the joint active and passive beamforming problem in IRS-assisted networks. In the following, we first review recent ML applications for IRS-assisted wireless networks. We then introduce our optimization-driven DRL framework that integrates the strengths of model-based optimization and model-free learning to achieve efficient, robust, and scalable beamforming design for IRS-assisted wireless networks.

2.4.2 DRL for IRS-Assisted Wireless Networks

DRL offers an interactive learning framework in which an agent learns to make sequential decisions by continuously interacting with a dynamic and often partially observable environment, receiving feedback in the form of immediate rewards or penalties. Such sequential decision making problems are mathematically modeled into MDP, defined by a set of system states, a set of actions, a state transition model, and a reward function. DRL employs DNNs as the parameterized approximators for the value functions, policies, and transition models, enabling the agent to handle high-dimensional and dynamic control problems common in wireless networks. By balancing exploration and exploitation, DRL seeks to discover an optimal policy that maximizes the expected long-term cumulative reward. In literature, DRL has demonstrated significant potential for autonomous decision-making in IRS-assisted networks, where it can efficiently optimize the IRS's passive beamforming and other coupled control variables even under dynamic and uncertain network conditions. Besides RL and DRL, supervised and unsupervised learning have also been extensively employed in IRS-assisted wireless networks for addressing the challenges associated with the high dimensionality and non-convexity of control problems, such as hybrid beamforming, user scheduling, and channel estimation. These ML approaches offer promising solutions by learning directly from data or environmental interactions, allowing systems to adapt to complex and uncertain scenarios without requiring exact analytical models. In the sequel, we provide a brief classification of representative applications of ML approaches in IRS-assisted wireless networks.

2.4.2.1 Supervised Learning for IRS-Assisted Channel Estimation

In IRS-assisted wireless networks, accurate channel information is critical for effective beamforming optimization [8]. A common approach to channel estimation involves a training phase, where known pilot signals are transmitted to the receiver and the received signals are used to infer channel conditions. This process can be framed as a supervised learning problem, where the input (pilot signal and received response) and the target output (channel matrix or parameters) form labeled training data. Typically, channel estimation is performed by activating one IRS element at a time while keeping the others inactive, isolating the contribution of each scattering element to the end-to-end channel. However, this approach incurs significant training overhead, especially for a large-size IRS, making it impractical in dynamic environments. To address this challenge, deep learning models have been proposed to learn the mapping from received signals to channel parameters. For instance, the authors in [9] employed a convolutional neural network (CNN) to estimate both the direct and cascaded (transmitter-IRS-receiver) channels in an IRS-assisted millimeter-wave (mmWave) system. The CNN was trained on simulated data pairs consisting of input pilot signals and corresponding ground-truth channel vectors, enabling real-time channel prediction after training. Channel estimation was modeled as a denoising problem in [10] and a deep residual network was proposed to implicitly learn the residual noise for recovering the channel coefficients from the noisy observations. The well-trained deep residual network is then used to identify the channel matrix given any noisy observation in the online phase. To further reduce training overhead and complexity, a hybrid IRS architecture was proposed in [11], where a sparse active element configuration is adopted and only a subset of IRS elements are equipped with sensing capabilities. A DNN was trained to reconstruct the full channel from partial observations, effectively denoising and interpolating the incomplete measurements. This approach significantly reduces pilot overhead while maintaining estimation accuracy in mmWave systems. A conditional generative adversarial network (CGAN) was proposed in [12] to estimate the cascaded channels. The received signals are the conditional input to the generator and discriminator, which are competing with each other to obtain an adaptive loss function and make more accurate channel estimations.

2.4.2.2 Supervised Learning for Beamforming Optimization

DNNs can be employed in a supervised learning framework to learn the mapping from environmental observations to optimal beamforming configurations. A dataset is first generated offline based on model-based optimization or large-scale system simulations, where input features including channel information, user location, or received pilot signals are paired with corresponding optimal beamforming actions (e.g., IRS's phase shifts or transmit precoders) that maximize a desired objective, such as received signal strength or spectral efficiency. The well-trained DNNs can be deployed for real-time inference, enabling fast adaptation to changing conditions without solving complex optimization problems online. For example, the authors in [2] trained a DNN to map a user's location and channel response to the IRS's optimal phase matrix, significantly reducing online computational latency. An appropriately trained CNN was employed in [13] to map the incident signals to the interfering user set, which can be used to control the binary statuses of the IRS elements that maximize the received signal to interference plus noise ratio (SINR). The authors in [14] adopted a fully-connected DNN to achieve the direct mapping from the partial channels to the optimal beamforming vector with maximal transmission rate. Similarly, graph neural networks (GNNs) were leveraged in [15] to enhance scalability and generalization in multi-user or large-scale IRS deployments. By modeling the network topology as a graph, GNNs can effectively capture spatial relationships and adapt to varying numbers of users or IRS elements, making them well-suited for dynamic network conditions.

2.4.2.3 Unsupervised Learning for Beamforming Optimization

Labeled training data can be costly or impractical to generate. Unsupervised learning approaches have also been proposed for beamforming optimization to eliminate the need for labeled training data. These methods directly optimize a performance-driven objective, e.g., signal-to-noise ratio (SNR), sum-rate, or energy efficiency, by designing a suitable loss function that can be minimized using gradient-based learning methods, without requiring ground-truth beamforming solutions. For instance, the authors in [3] developed an unsupervised DNN framework for optimizing IRS's beamforming, where the loss function is constructed based on the negative SNR. The DNN framework learns to adapt the IRS's phase shifts to effectively maximize the desired signal strength without prior knowledge of the optimal beamforming configurations. This approach offers greater flexibility and adaptability to real-time channel variations and hardware imperfections, avoiding the overhead of dataset generation and labeling.

2.4.2.4 DRL for Beamforming Optimization

DRL offers a powerful framework for online beamforming optimization in IRS-assisted networks by continuous interaction with the dynamic wireless environment. Unlike supervised or unsupervised learning, DRL does not require labeled datasets or explicit channel models, making it well-suited for real-time and model-free decision-making under uncertainty. When the IRS's phase control is discretized into a finite set of actions, DQN can be employed to select the optimal phase configuration based on observed channel conditions and feedback from the receiver, e.g., signal strength or data rate [16]. The agent learns a value function that maps state-action pairs to expected long-term rewards, enabling efficient exploration of the action space and convergence to high-performing beamforming strategies. Similarly, DDQN was employed in [17] to perform real-time coarse phase control for IRS. For more precise control with the IRS's continuous phase shifts, DDPG are preferred to enable a direct optimization of the continuous phase vector that maximizes a desired performance metric, such as the received SNR in an IRS-assisted MISO system [18] and the sum rate in a multi-user scenario [19]. Moreover, DRL can enhance physical layer security in IRS-assisted networks. The authors in [20] leveraged DDPG to jointly optimize transmit beamforming and IRS's phase shifts, steering the signal toward legitimate users while minimizing information leakage to eavesdroppers.

Compared to conventional optimization-based approaches, ML approaches offer enhanced flexibility and robustness in IRS-assisted wireless networks. They are inherently resilient to uncertain channel information, imperfect modeling assumptions, and rapidly changing environmental conditions. Unlike model-based optimization,

ML approaches learn implicit patterns from data and generalize across diverse scenarios. This enables adaptive decision-making without requiring explicit problem formulation, making them particularly suitable for real-time control in complex and dynamic networks. Despite their advantages, the practical deployment of ML approaches in IRS-assisted networks still faces several critical challenges, mainly due to the requirement for an offline training or slow convergence in online learning. Supervised and unsupervised learning methods typically require extensive offline training on large datasets. These datasets are often generated using simplified or simulated channel models, which may not fully capture real-world complexities such as hardware impairments, non-ideal reflections, or mobility effects. As a result, the trained models may suffer from systematic bias when deployed in practical environments, leading to suboptimal performance during online inference. While DQN and DDPG methods can learn optimal policies from scratch through environment interaction, they often exhibit slow convergence due to the high dimensionality of the state and action spaces. In IRS-assisted networks, the large number of reflecting elements leads to an exponentially large action space, making it inefficient and time-consuming for action exploration during training. This is particularly problematic in highly dynamic network environments where rapid adaptation is required, and prolonged online learning may incur significant performance loss.

2.4.3 Optimization-Driven DDPG for Joint Beamforming

Our case study in this section investigates the application of the optimization-driven DRL framework in an IRS-assisted multi-input single-output (MISO) downlink communication system [4]. As illustrated in Fig. 2.3, a multi-antenna AP transmits information to single-antenna receivers with the aid of an IRS equipped with N reflecting elements. We assume that the IRS can adjust both the continuous phase shift and the reflection magnitude of each element, enabling fine-grained control over the reflected RF signals. While continuous phase and amplitude control is considered, the framework can be readily extended to practical scenarios with discrete phase shifts by constraining the feasible set accordingly. The IRS operates in a self-sustainable manner by harvesting RF energy from the incident signals. This is achieved through a power-splitting (PS) mechanism, where a portion of the incident signal is reflected toward the receiver and the remainder is harvested and converted into energy to power the IRS's operations. Channel state information (CSI) is subject to uncertainty due to estimation errors by using the passive IRS. We assume that the average channel conditions can be obtained through measurements. The estimation errors are modeled as random variables distributed in a bounded and convex set.

Our objective is to minimize the AP's transmit power through a joint active and passive beamforming optimization, subject to the IRS's energy constraint and the receiver's SNR requirement. The problem formulation follows a similar structure to that in [21]. As illustrated in Fig. 2.3, the overall control variables include the AP's active beamforming vector $\mathbf{w}$, the IRS's PS ratio ρ to balance energy harvesting and signal enhancement, and the IRS's phase shift vector $\boldsymbol{\theta}$ to configure the reflected channel conditions. This joint beamforming optimization is inherently challenging due to the non-convex coupling between the AP's active and the IRS's passive beamforming strategies. Traditional methods such as alternating optimization (AO) suffer from high computational complexity and slow convergence, especially in the presence of uncertain CSI. To address these challenges, we employ the optimization-driven DDPG framework. At each decision epoch t, the action is defined as $\mathbf{a}_t = (\rho_t, \mathbf{w}_t, \boldsymbol{\theta}_t)$, which can be decomposed into two parts, i.e., the scalar PS ratio ρ_t and the beamforming pair $(\mathbf{w}_t, \boldsymbol{\theta}_t)$. Here we adopt a hierarchical decomposition to reduce complexity and improve learning efficiency. In the outer-loop DDPG, the PS ratio ρ_t is first generated by DDPG's actor network. Given fixed ρ_t, we can determine a feasible phase vector $\boldsymbol{\theta}_t$ using simple heuristic rules (e.g., aligning phases with channel angles). Subsequently, the optimal active beamforming vector $\mathbf{w}_t$ can be obtained efficiently by solving an inner-loop convex optimization subproblem, which significantly reduces computational burden.

As shown in Fig. 2.1d, DDPG's actor and critic networks independently generate action proposals and value estimates. The proposed action ρ_t is then passed to an

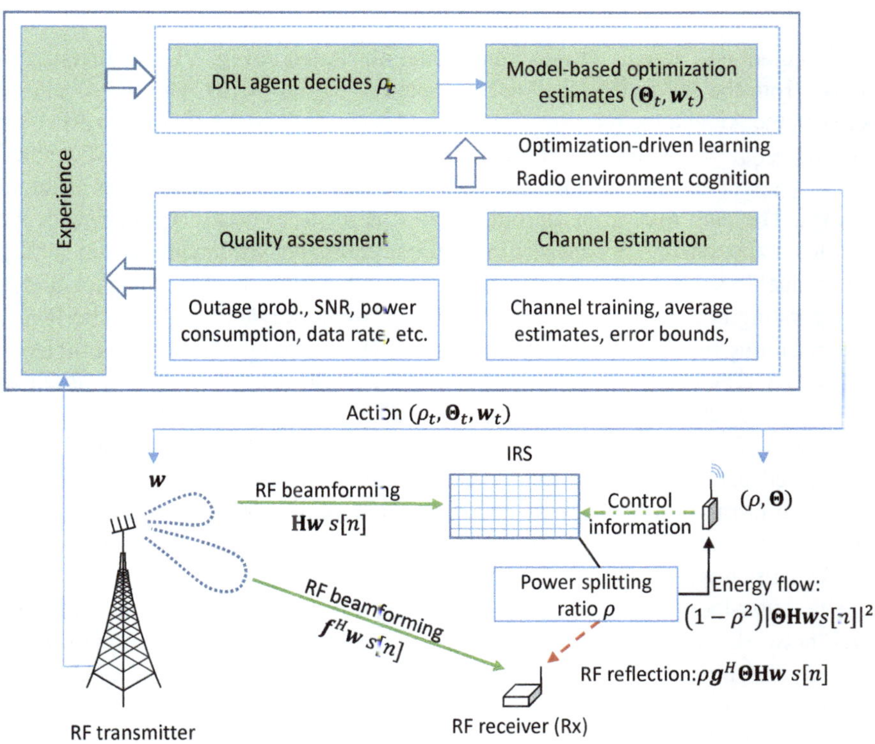

Fig. 2.3 Case study: Optimization-driven DRL for energy-efficient beamforming optimization in a self-sustainable IRS-assisted MISO downlink wireless communication system

optimization module, which computes the optimized solution $(\mathbf{w}_t, \boldsymbol{\theta}_t)$ and evaluates a lower bound on the target Q-value based on the resulting system performance. If the optimization-driven target Q-value exceeds the estimate from the target Q-network of the outer-loop DDPG, it will be used with a higher probability as the new target Q-value for DNN training of the outer-loop DDPG's critic network. Simultaneously, the action $\mathbf{a}_t$ is updated with the optimized solution $(\mathbf{w}_t, \boldsymbol{\theta}_t)$, effectively injecting analytical knowledge into the learning process. An alternative decomposition of control variables is also feasible. Since the IRS's phase vector $\boldsymbol{\theta}_t$ is difficult to optimize directly due to its high dimensionality and non-convex couplings, it can be searched by the outer-loop DDPG, while the remaining variables $(\rho_t, \mathbf{w}_t)$ are optimized jointly via a line search over ρ_t for each outer-loop action $\boldsymbol{\theta}_t$. This alternative approach reduces the outer-loop DDPG's action space and leverages convexity of the inner-loop optimization to accelerate convergence. Compared to conventional model-free DDPG, this hybrid optimization-driven DDPG framework significantly enhances learning stability, convergence speed, and solution quality.

2.5 Numerical Evaluation

The simulation is based on the system model illustrated in Fig. 2.3. The distance in meters from the AP to the receiver is denoted as $d_{\text{AP,User}} = 20$. The IRS is placed between the AP and the receiver. The vertical distance from the IRS to the AP-User link is $d_{\text{IRS,User}} = 5$ meters. The horizontal distance from the AP to the IRS is denoted by $d_{\text{AP, IRS}}$, which varies during simulations to evaluate performance under different settings. The path loss at the unit distance is assumed to be $L_0 = 30$ dB with the path-loss exponent of 3.5, consistent with urban cellular environments as that in [22]. All channel links (AP-IRS, IRS-User, and AP-User) follow the same channel model, incorporating both large-scale fading and small-scale Rayleigh fading. The IRS is self-sustaining through RF energy harvesting, with an energy conversion efficiency of $\eta = 0.5$. The receiver's noise power is set to -80 dBm. The IRS is equipped with $N = 20$ reflecting elements, and the AP is equipped with $M = 2$ antennas. The reward for the outer-loop DRL agent is defined as the ratio of successfully transmitted data to the AP's total energy consumption, which promotes energy-efficient transmission. This reward encourages the agent to minimize transmit power while ensuring reliable communication under the receiver's SNR constraint and the IRS's energy requirement.

The outer-loop DDPG employs fully connected DNNs for both actor and critic networks. Let `fc(m,n)` denote a fully connected layer with input dimension m and output dimension n. The critic network takes concatenated state and action inputs through separate fully connected layers (`fc(N_STATES,64)` and `fc(N_ACTIONS,64)`), which are then combined, passed through a `relu` activation, and mapped to a scalar output via `fc(64,1)`. Here `N_STATES` and `N_ACTIONS` specify the dimensions of the state and action spaces, respectively. The actor network consists of two layers `fc(N_STATES,64)` and `fc(64,N_ACTIONS)`, with a `sigmoid` activation to produce bounded actions. The algorithm uses a batch size of 32, critic and actor learning rates of 1×10^{-4} and 1×10^{-3}, respectively, a reward discount factor of 0.5, a replay memory capacity of 1000, and a soft update factor of 0.01 for target network stabilization. For the given configuration, i.e., $M = 2$ antennas and $N = 20$ reflecting elements, the control variables include the following parts: the scalar PS ratio ρ_t, the phase shift vector $\boldsymbol{\theta}_t$ with 40 real-valued parameters and active beamforming vector $\mathbf{w}_t$ with 4 real parameters (real and imaginary parts). Thus, the full action space has dimension of 45. To ensure a fair comparison, both the optimization-driven DDPG and the model-free DDPG use the same DNN architecture and hyperparameters. The key difference lies in the action space dimensionality. In model-free DDPG, the agent learns all 45 control variables directly. In optimization-driven DDPG, only a subset of variables (e.g., the PS ratio ρ_t, or the phase shift vector $\boldsymbol{\theta}_t$) is learned by the DNN. The remaining variables are optimized via embedded convex solvers or line searches. This splitting of control variables significantly reduces the effective action space size `N_ACTIONS`, leading to faster convergence and improved reward performance.

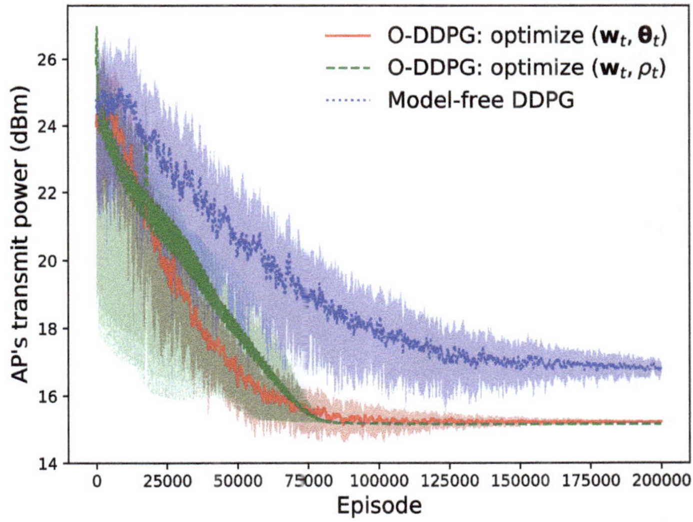

Fig. 2.4 AP's transmit power under different DDPG algorithms. The solid line represents the median across 50 independent trials, while the shaded regions indicate the 10th to 90th percentiles for each algorithm. O-DDPG uses inner-loop optimization, either joint heuristic or bisection-based, to refine beamforming and power splitting strategies. This hierarchical learning approach enables faster convergence and lower transmit power by improving action selection and Q-value estimation, demonstrating superior efficiency and performance over model-free DDPG in transmit power minimization

2.5.1 *Improved Reward Performance*

Figure 2.4 illustrates the evolution of the AP's transmit power when comparing the optimization-driven DDPG (O-DDPG) with the conventional model-free DDPG. In the O-DDPG framework, we consider two different approaches for the inner-loop optimization. In the first case, given the power splitting ratio ρ_t determined by the outer-loop DDPG, we jointly optimize the active beamforming vector $\mathbf{w}_t$ and the passive beamforming matrix $\boldsymbol{\theta}_t$ using a heuristic optimization method. In the second case, the passive beamforming $\boldsymbol{\theta}_t$ is selected by the outer-loop DDPG, while $(\rho_t, \mathbf{w}_t)$ are optimized via a bisection-based method. In both cases, the solutions from the inner-loop optimization are used to refine the outer-loop action selection and provide more accurate Q-value estimates. By leveraging this hierarchical structure, O-DDPG achieves faster convergence and significantly reduces the AP's transmit power compared to the model-free DDPG.

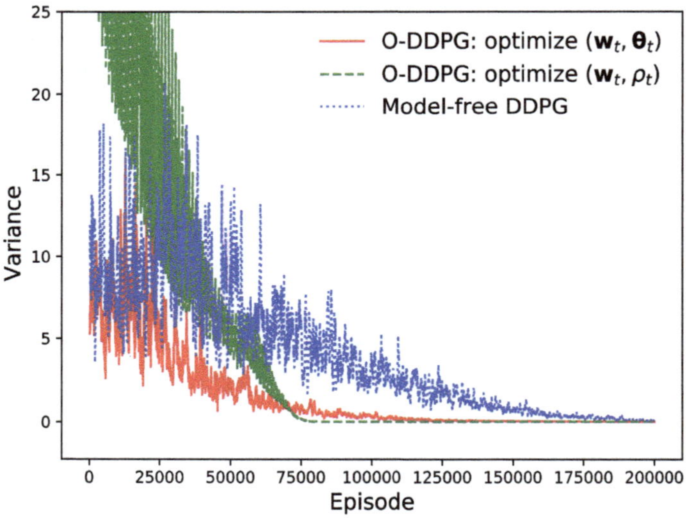

Fig. 2.5 O-DDPG achieves more stable learning dynamics and faster convergence compared to model-free DDPG. O-DDPG stabilizes by 10^6 episodes, while model-free DDPG requires approximately 75% more episodes. This demonstrates that the optimization-driven framework enhances learning stability, reduces fluctuations, and accelerates convergence, leading to more efficient and robust training in wireless network control. The system geometry is defined by the following distances: the AP-User distance $d_{\text{AP,User}} = 20$, the AP-IRS horizontal distance $d_{\text{AP,IRS}} = 10$, and the vertical distance from the IRS to the AP-User link $d_{\text{IRS,User}} = 5$

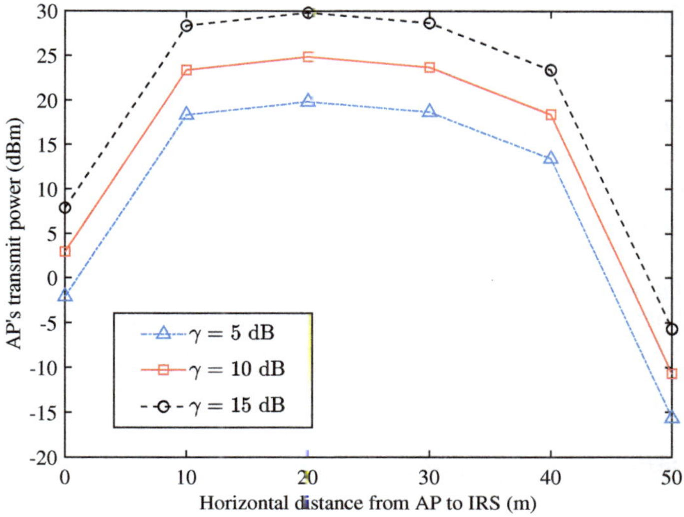

Fig. 2.6 The AP's transmit power reaches its maximum when IRS is positioned midway between the AP and the receiver. In the ideal case, with negligible IRS's power consumption, the AP reduces its transmit power as IRS moves to the receiver. The AP-User distance is $d_{\mathrm{AP,User}} = 50$ m

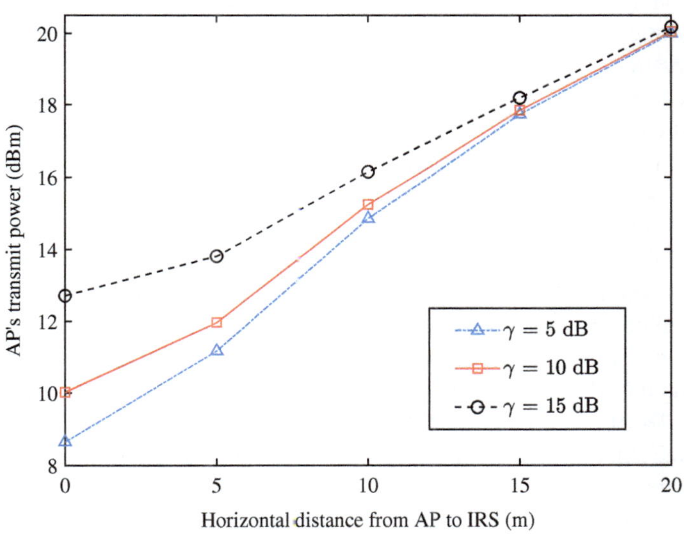

Fig. 2.7 In practical scenarios with non-zero IRS's power consumption, the AP increases its transmit power as the IRS moves away. This compensates for weaker channel gains, ensuring sufficient RF energy for IRS's energy harvesting. As more signal is allocated to energy harvesting, less power remains for reflection and information transmission. The IRS's power demand is set as a constant of 20 μW, which is sustained by RF energy harvested from the wireless environment. The AP-User distance is $d_{\mathrm{AP,User}} = 20$ m

2.5.2 Faster and Stable Learning

From Fig. 2.4, we also observe that the shaded regions (representing reward variation) for the O-DDPG algorithms are narrower than those of the model-free DDPG, indicating more stable learning performance. To quantify this stability, we evaluate the reward's variance across learning episodes. As shown in Fig. 2.5, both O-DDPG variants exhibit significantly lower reward variance compared to the model-free DDPG. Furthermore, the O-DDPG algorithms converge earlier, with reward variance approaching zero within fewer learning episodes. Specifically, the O-DDPG algorithms stabilize after 10^6 episodes, whereas the model-free DDPG requires approximately 75% more episodes to achieve comparable stability.

2.5.3 Optimal Deployment Location

Figures 2.6 and 2.7 depict the AP's transmit power as the IRS moves from the AP toward the receiver. In the ideal case, where the IRS's power consumption is negligible, the IRS remains fully active and consistently enhances the AP's information transmission. As shown in Fig. 2.6, for a fixed SNR requirement at the receiver, the AP can significantly reduce its transmit power as the IRS moves closer to the receiver. This observation aligns with previous findings in [22], which suggest that deploying the IRS near the receiver is more beneficial under ideal assumptions. However, in a more practical scenario where the IRS has a non-zero power demand, the behavior changes significantly. As illustrated in Fig. 2.7, the AP must increase its transmit power to ensure that sufficient RF energy is available for the IRS to harvest and sustain its operations as it moves away from the AP. This is because a portion of the transmitted RF signal is allocated to energy harvesting at the IRS by controlling its PS ratio. As the AP-IRS distance increases, the channel gain weakens, prompting the IRS to allocate a larger fraction of the received signal to energy harvesting. Consequently, less signal power is available for reflection and information transmission, diminishing the IRS's effectiveness.

This trade-off reveals a critical design insight. When the IRS has non-negligible power requirements, placing it too far from the transmitter undermines its ability to assist wireless communications. Therefore, the optimal IRS's deployment must balance proximity to the receiver for better end-to-end channel quality and proximity to the transmitter for more reliable energy harvesting. This contrasts with the conventional observation in [22] and highlights that the IRS's power demand is a key factor to determine its optimal placement in practical wireless networks.

2.6 Discussions and Conclusions

In this chapter, we have reviewed DRL's applications in complex wireless networks. By examining common limitations of existing DRL approaches, we have been motivated to design a novel optimization-driven DRL framework. This framework leverages partial domain knowledge and model-based optimization to guide the learning process, addressing key challenges in scalability and convergence. The effectiveness of the proposed framework is demonstrated through a case study in an IRS-assisted wireless network. Numerical results show that, compared to conventional model-free DRL methods, the optimization-driven framework achieves significantly improved learning efficiency and reward performance. These results confirm that integrating model-based optimization with model-free DRL can enhance decision-making in complex environments by providing more informed policy updates and value estimations. Looking ahead, several promising directions can further advance the applications of optimization-driven DRL in wireless networks:

- Control variable splitting: The optimization-driven DRL framework divides the control variables into those handled by outer-loop learning and those optimized in the inner loop. The effectiveness of this decomposition depends on striking a balance between the computational complexity of the inner-loop optimization and the convergence speed of the outer-loop learning process. However, there are currently no universal guidelines for determining an optimal variable splitting strategy. For the same control problem, different partitioning schemes can lead to significantly divergent learning performance. Therefore, a key direction for future research is to develop a systematic and analytical framework to evaluate and quantify the quality of various variable splitting strategies, enabling principled design choices for the optimization-driven DRL framework.
- Performance trade-off: The optimization-driven DRL framework achieves faster convergence by reducing the action space through inner-loop optimization. However, the overall reward may be compromised, particularly when a subset of control variables is governed by imperfect or simplified physical models and optimized approximately. In such cases, suboptimal solutions in the inner loop can propagate to the outer loop, limiting the achievable performance. This introduces a fundamental trade-off between learning efficiency and asymptotic reward performance. Understanding and quantifying this trade-off is crucial for designing effective hybrid learning framework integrating model-based optimization with model-free DRL. Therefore, it is worth exploring this balance systematically in the future work, as it can provide principled guidance for the optimal splitting of control variables in the optimization-driven DRL framework.
- Adaptive integration of optimization: The model-based optimization significantly enhances learning efficiency in the early stages of training, but its marginal contribution diminishes as the agent accumulates experience and the reward approaches convergence. Initially, when the DRL agent lacks sufficient historical data and explores largely at random, actions derived from inner-loop optimization tend to yield higher rewards and are thus more frequently selected. This naturally pro-

motes a tighter integration of optimization and learning at the beginning of the training process. As learning progresses and the policy stabilizes, the benefit of frequent optimization decreases. This suggests an opportunity for adaptive integration of the model-based optimization into the model-free learning. Instead of executing the computationally intensive optimization module in every decision epoch, it should be invoked only when necessary, such as when the outer-loop variables or the reward performance undergo significant changes. Such an on-demand adaptive integration strategy can reduce computational overhead while preserving the benefits of model-based optimization, leading to a more efficient and scalable learning framework.

- Optimization-driven multi-agent learning: The case study presented in this chapter demonstrates the feasibility and advantages of integrating model-based optimization with model-free DRL. This motivates further exploration into other optimization-driven ML paradigms, as well as hybrid approaches combining multiple such design strategies. One promising direction is the extension of the optimization-driven DRL framework to multi-agent systems, where each agent makes decisions based on observations and actions of other agents. In such settings, conventional multi-agent DRL methods often suffer from high sample complexity and slow convergence due to the exponentially enlarged joint action and state spaces. To address these challenges, agents can leverage local model-based optimization to estimate or predict the behaviors of neighboring agents, based on shared or inferred system models. These predictions can then guide policy updates and value estimation, effectively reducing action exploration burden and accelerating convergence. Such synergy between local model-based optimization and distributed learning holds strong potential for scalable and efficient coordination in complex wireless networks.

References

1. A. Zappone, M. Di Renzo, M. Debbah, T.T. Lam, X. Qian, Model-aided wireless artificial intelligence: embedding expert knowledge in deep neural networks for wireless system optimization. IEEE Veh. Techn. Mag. **14**(3), 60–69 (2019)
2. A. Taha, M. Alrabeiah, A. Alkhateeb, Deep learning for large intelligent surfaces in millimeter wave and massive MIMO systems, in *Proceedings of IEEE GLOBECOM*, pp. 1–6 (2019)
3. J. Gao, C. Zhong, X. Chen, H. Lin, Z. Zhang, Unsupervised learning for passive beamforming. IEEE Commun. Lett. **24**(5), 1052–1056 (2020)
4. S. Gong, J. Lin, B. Ding, D. Niyato, D.I. Kim, M. Guizani, When optimization meets machine learning: the case of IRS-assisted wireless networks. IEEE Network **36**(2), 190–198 (2022)
5. S. Gong, X. Lu, D. T. Hoang, D. Niyato, L. Shu, D. I. Kim, Y. Liang, Towards smart wireless communications via intelligent reflecting surfaces: a contemporary survey. IEEE Commun. Surv. Tut. **22**(4), 1–33 (2020)
6. D. Muirhead, M.A. Imran, K. Arshad, A survey of the challenges, opportunities and use of multiple antennas in current and future 5G small cell base stations. IEEE Access **4**, 2952–2964 (2016)

7. M.D. Alessio Zappone, M. Di Renzo, Wireless networks design in the era of deep learning: model-based, AI-based, or both? IEEE Trans. Commun. **67**(10), 7331–7376 (2019)
8. B. Zheng, C. You, W. Mei, R. Zhang, A survey on channel estimation and practical passive beamforming design for intelligent reflecting surface aided wireless communications. IEEE Commun. Surv. Tut. **24**(2), 1035–1071 (2022)
9. A.M. Elbir, A. Papazafeiropoulos, P. Kourtessis, S. Chatzinotas, Deep channel learning for large intelligent surfaces aided mm-Wave massive MIMO systems. IEEE Wireless Commun. Lett. **9**(9), 1447–1451 (2020)
10. C. Liu, X. Liu, D.W.K. Ng, J. Yuan, Deep residual learning for channel estimation in intelligent reflecting surface-assisted multi-user communications. IEEE Trans. Wireless Commun. **21**(2), 898–912 (2022)
11. S. Liu, Z. Gao, J. Zhang, M.D. Renzo, M.-S. Alouini, Deep denoising neural network assisted compressive channel estimation for mmWave intelligent reflecting surfaces. IEEE Trans. Veh. Techn. **69**(8), 9223–9228 (2020)
12. M. Ye, H. Zhang, J.-B. Wang, Channel estimation for intelligent reflecting surface aided wireless communications using conditional GAN. IEEE Commun. Lett. **26**(10), 2340–2344 (2022)
13. B. Yang, X. Cao, C. Huang, C. Yuen, L. Qian, M.D. Renzo, Intelligent spectrum learning for wireless networks with reconfigurable intelligent surfaces. IEEE Trans. Veh. Technol. **70**(4), 3920–3925 (2021)
14. S. Zhang, S. Zhang, F. Gao, J. Ma, O.A. Dobre, Deep learning optimized sparse antenna activation for reconfigurable intelligent surface assisted communication. IEEE Trans. Commun. **69**(10), 6691–6705 (2021)
15. T. Jiang, H.V. Cheng, W. Yu, Learning to reflect and to beamform for intelligent reflecting surface with implicit channel estimation. IEEE J. Sel. Area. Commun. **39**(7), 1931–1945 (2021)
16. G. Lee, M. Jung, A.T.Z. Kasgari, W. Saad, M. Bennis, Deep reinforcement learning for energy-efficient networking with reconfigurable intelligent surfaces, in *Proceedings of IEEE ICC*, pp. 1–6 (2020)
17. W. Wang, W. Zhang, Intelligent reflecting surface configurations for smart radio using deep reinforcement learning. IEEE J. Selected Area. Commun. **40**(8), 2335–2346 (2022)
18. K. Feng, Q. Wang, X. Li, C. Wen, Deep reinforcement learning based intelligent reflecting surface optimization for MISO communication systems. IEEE Wireless Commun. Lett. **9**(5), 745–749 (2020)
19. C. Huang, R. Mo, C. Yuen, Reconfigurable intelligent surface assisted multiuser MISO systems exploiting deep reinforcement learning. IEEE J Sel. Area. Commun. **38**(8), 1839–1850 (2020)
20. H. Yang, Z. Xiong, J. Zhao, D. Niyato, L. Xiao, Q. Wu, Deep reinforcement learning based intelligent reflecting surface for secure wireless communications. IEEE Trans. Wireless Commun. **20**(1), 1963–1974 (2021)
21. J. Lin, Y. Zou, X. Dong, S. Gong, H. T. Dinh, D. Niyato, Deep reinforcement learning for robust beamforming in IRS-assisted wireless communications, in *Proceedings of IEEE GLOBECOM*, pp. 1–6 (2020)
22. Q. Wu, R. Zhang, Intelligent reflecting surface enhanced wireless network via joint active and passive beamforming. IEEE Trans. Wireless Commun. **18**(11), 5394–5409 (2019)

Chapter 3
Hierarchical DRL for Heterogeneous Wireless Networks

3.1 Introduction

Conventionally, network optimization methods are typically based on simplified system models to be tractable efficiently, e.g., with perfect channel information or ideal system implementation. The optimization solutions based on simplified system models only provide lower bounds or approximations to the actual system performance. In practice, wireless networks normally contain diverse entities, such as centralized controllers, distributed base stations, and end-user devices, with heterogeneous capabilities in computation, memory, and energy resources. Such a heterogeneity significantly complicates the network control problems, which often involve high-dimensional and mixed-integer control variables, e.g., including beamforming optimization, access control, transmit scheduling, relay selection, etc. This inherently requires decentralized coordination among all different entities to ensure scalability and flexibility. Each network entity independently manages its own local resource allocation while all devices collaboratively contribute to the overall network performance through information sharing and dynamic coordination.

The distributed nature of wireless networks calls for a heterogeneous DRL framework that can decompose the network optimization problem into different tasks and assign the control tasks to different network entities. The basic idea is to distribute different control variables across network entities using a multi-level learning structure. A high-level DRL agent can be deployed at the network controller to handle global decision-making, learning policies that are resilient to model uncertainties and dynamic environment. The low-level local devices employ optimization-based techniques to solve specific control tasks, such as adjusting transmit power or scheduling decisions based on the policies and guidance received from the upper level. Given the observation of network status, The high-level network controller and the low-level end user can update their local control variables in an iterative manner. The interaction and iterative control between different entities can help separate different

S. Gong et al., *Optimization-Driven Deep Reinforcement Learning for Wireless Networks*, Wireless Networks, https://doi.org/10.1007/978-3-032-22997-7_3

control variables and reduce the search space for each entity. Such a hierarchical structure offers several key advantages as follows:

- By decoupling control tasks in different network entities, the hierarchical framework reduces dimensionality of action space for each agent, thereby accelerating convergence of learning and improving overall scalability.
- The integration of local model-based optimization into the outer-loop DRL allows for more informed and efficient exploration of control variables, especially when partial model information is available.
- The iterative coordination between network entities provides more flexibility that enables the system to dynamically adapt to changing network conditions and user requirements.

In this chapter, we propose the hierarchical DRL framework for network optimization in complex wireless networks with both discrete and continuous variables, by exploiting the efficiency of optimization methods and the robustness of DRL approaches. In particular, we use model-free DRL to build the learning framework at the network controller that is robust to complex problem structure and inexact modeling. Meanwhile we integrate local optimization methods for wireless users in the inner loop to reduce the search space and improve the learning efficiency. The following case study and detailed algorithm design were originally proposed in our earlier work [1]. This chapter refines and reorganizes the content from [1] to ensure consistency, clarity, and coherence within the broader scope of the book.

3.2 Case Study in Backscatter-Aided Relay Communications

Our case study in this chapter focuses on throughput maximization in a backscatter-aided relay communications system, considering a novel relaying strategy that allows each relay node to switch on-demand between low-power backscatter communications and the active RF communications. Backscatter communications technology has recently emerged as a promising technology to enhance both the energy- and spectrum-efficiency of future wireless networks, e.g., [2, 3]. Unlike conventional RF communications, which require active components and significant energy consumption for signal transmission and reception, backscatter radios operate in a passive mode by modulating and reflecting incident RF signals through controllable variations in their antenna's matching impedance. This allows the backscatter radios to communicate without generating carrier signals by themselves, significantly reducing power consumption. As such, the backscatter radios are expected to be self-sustained by harvesting RF energy from ambient energy sources, making them well-suited for low-power or battery-free IoT applications, e.g., [4, 5].

Backscatter communications typically have a low data rate and suffer from vulnerability to channel variations. The backscatter radios' passive reflections make it highly susceptible to environmental changes, multipath fading, and signal attenuation. Due to simplified circuitry implementation and limited signal processing capability, backscatter radios typically achieve lower SNR and are more vulnerable to interference. Hence, it is more preferable to take advantages of both RF and backscatter communications in a hybrid radio network, where hybrid radios can dynamically switch between the active and passive modes according to their current energy statuses, traffic demands, and channel conditions, e.g., [6, 7]. Such a hybrid design enables a more flexible and adaptive communication strategy. For example, hybrid relay nodes equipped with both active and backscatter capabilities can extend network coverage and improve connectivity for energy-constrained devices. When the channel conditions are favorable or the energy supply is limited, a relay can choose to operate in the backscatter mode, reflecting and modulating signals from the other nodes. Conversely, under poor link quality, the node may switch to the active mode to ensure more reliable and robust communications.

The hybrid radios' dynamic switching between different communication modes also introduces a new dimension to the network control problem, which involves the joint optimization of mode selection, relay scheduling, beamforming, and power allocation under time-varying channel conditions and energy constraints. To address these challenges, DRL is a powerful tools to learn optimal mode selection and resource control policies without requiring explicit modeling of channel distributions or energy harvesting processes. DRL agent can be trained to observe the environment, such as channel quality, energy level, and traffic demand, and then make intelligent decisions on whether to activate backscatter radios, which relay to select, and how to allocate transmit power. This enables the system to continuously adapt to dynamic environment while minimizing energy consumption and maintaining communication quality. By integrating passive and active radios, we can achieve an efficient balance between energy savings and transmission performance.

Generally, there are two relaying strategies that can be employed to exploit the complementary operations of hybrid radios. The first strategy follows a decode-and-forward (DF) approach. In particular, the hybrid relay receives and decodes the information signal from the source in the first hop. Subsequently, it encodes and forwards the message to the destination using either active RF or backscatter communication, depending on energy availability and channel conditions. This DF-alike scheme enables a more flexible relaying, where high-rate active RF transmission can be used when energy is sufficient, while backscatter communication mode offers an ultra-low-power alternative for energy-constrained scenarios. However, this approach incurs processing delay and is susceptible to error propagation, i.e., the decoding failure in the first hop will result in retransmissions and degrading end-to-end reliability. The second strategy is analogous to the conventional amplify-and-forward (AF) scheme, but implemented via passive backscattering. In this case, the relay does not decode the signal, but instead it instantaneously reflects and modulates the incident RF signal toward the destination by adjusting its reflection coefficient [7]. Since no decoding or active RF transmission is required, the passive relay operates with minimal power consumption, making it highly energy-efficient and well-suited for battery-free IoT scenarios. Numerical results in [8] showed that the RF signal detection performance can be improved even with a single AF-alike passive relay. Furthermore, the bit-error-rate (BER) analysis in [9] revealed that both schemes perform similarly under ideal conditions. However, under more practical operating conditions, e.g., with imperfect channel information, decoding errors, and timing mismatches, the DF-alike scheme suffers significant performance degradation due to error propagation and processing latency, whereas the AF-alike scheme remains robust and often outperforms the DF-alike relay scheme. These results highlight the potential of AF-alike relay scheme as a low-complexity, low-latency, and energy-efficient solution for enhancing wireless connectivity.

The case study in this chapter aims to enhance system throughput by leveraging both radio diversity gain and multi-user cooperation gain in a hybrid relay-assisted wireless network. We consider a more complex and practical scenario where multiple hybrid radio relays capable of operating in either active RF or passive backscatter mode cooperatively assist the transmission from a hybrid access point (HAP) to a destination receiver. Each relay is allowed to select its operational mode independently based on local energy availability and channel conditions. In contrast to prior works [8–15], which typically treat active and passive relay communications separately, we propose a hybrid relaying scheme in which both active relays and AF-alike passive relays simultaneously assist the same RF communication link. Specifically, the energy-constrained relays can operate in passive mode, reflecting and modulating the incident RF signals to improve the effective channel between the HAP and the receiver without consuming significant power. Whereas power-sufficient relays operate in the active RF mode, receiving and amplifying the signal before retransmitting it to enhance its reliability and data rate. The coexistence of heterogeneous relay schemes introduces a highly non-convex and combinatorial optimization problem, encompassing the radio's mode selection, the design of complex reflection coefficient, the energy harvesting scheduling, transmit power allocation, and hybrid beamforming strategies. Even when relay modes are fixed, the joint optimization of beamforming and relaying strategies remains challenging due to the strong coupling among multiple relays' decisions. Nonetheless, we envision that this hybrid relaying scheme offers enhanced flexibility and adaptability in energy-limited wireless networks. A similar system was studied in [16], where relays can choose their operating modes based on individuals' energy and channel states. The relays' optimal mode selection was formulated as a mixed-integer nonlinear program, which is computationally prohibitive for large-scale or real-time deployment. The case study in this chapter aims to develop more efficient, robust, and scalable solution that fully exploits the potential of hybrid relaying for energy-constrained wireless networks.

Compared to model-based optimization methods, model-free DRL has demonstrated superior flexibility and robustness in tackling complex and dynamic control problems in multi-user wireless networks, featuring imperfect system models, intricate user couplings, uncertain network dynamics, and high-dimensional decision spaces. For instance, the authors in [17] proposed an online RL framework to optimize spectrum access policies in ambient backscatter systems, accounting for time-varying channel conditions, energy availability, and traffic demands. DRL has further been applied to enhance security in hybrid radio networks, such as defending against intelligent jammers in [18] and [19]. Other works have successfully leveraged DRL for transmission scheduling [20], power allocation [21], and radio mode selection [22], demonstrating its versatility in network performance optimization.

In this chapter, we propose a novel hierarchical DDPG algorithm, denoted as H-DDPG, to address the challenges posed by high-dimensional and hybrid action spaces that combine discrete and continuous control variables. The case study aims to maximize system throughput in a hybrid relaying network by jointly optimizing relay mode selection, beamforming, and time allocation strategies. The first key design principle of H-DDPG is the decomposition of action space via a two-level hierarchical structure. Specifically, in the outer loop, DQN is employed responsible for making discrete mode selection decisions for each relay. In the inner loop, DDPG is then used to optimize the continuous control variables, including beamforming and time allocation for energy harvesting. This hierarchical decomposition significantly reduces policy complexity and enhances learning stability. The second innovation is the integration of model-based optimization into the model-free DDPG framework to provide better guidance and accelerate its training. Specifically, we incorporate analytical solutions and performance bounds derived from model-based optimization methods as informed targets for the outer-loop learning process, leading to faster convergence and improved reward performance. To summarize, the main technical contributions of this case study are as follows:

- We propose a multi-relay-assisted communication system in a hybrid radio network, in which each relay is capable of both active RF and passive communications, and can dynamically select its operating mode to assist the HAP's RF communications to the destined receiver. Passive relays enhance the effective RF channel conditions by backscattering the incident signals, while active relays utilize energy harvested from the HAP to amplify and forward the received signals, providing higher data rate and enhanced reliability. This dual-mode cooperation enables more flexible and energy-efficient relay communications for energy-constrained wireless networks.
- We formulate a throughput maximization problem to jointly optimize time allocation, transmit beamforming, and relaying strategies in the hybrid relay network. Considering the non-convexity and complexity arising from nonlinearly coupled variables, we propose a tractable approximation of the original problem, which is solved efficiently via monotonic optimization method by exploiting its structural properties. The approximate solution provides a lower bound on the throughput performance, which is used to guide the optimization of relays' operating modes.
- A hierarchical DRL framework, referred to as H-DDPG, is proposed to solve the throughput maximization problem. The outer-loop DQN is responsible for updating the relays' mode selection, while the inner-loop DDPG optimizes relays' reflection coefficients, transmit beamforming, and time allocation. We further enhance H-DDPG's outer-loop learning efficiency by leveraging the inner-loop performance lower bound to estimate a target Q-value, which is then integrated with DQN's training process to guide the relay mode selection. Numerical results

demonstrate that the optimization-driven H-DDPG achieves a higher cumulative reward and stable convergence compared to the conventional model-free DDPG.

3.3 System Model and Problem Formulation

We consider a wireless network consisting of a multi-antenna HAP and a group of single-antenna user devices, indexed by the set $\mathcal{N} = \{1, 2, \ldots, N\}$. These devices operate in a TS protocol, i.e., they first harvest RF energy from the HAP and subsequently assist in relaying information from the HAP to its intended receiver. The proposed system model can be naturally extended to scenarios with multiple receivers. Under a time division multiple access (TDMA) protocol, each receiver is allocated a dedicated time slot for relay-assisted transmission, enabling sequential and interference-free communication. The HAP is assumed to have a constant power supply and operates with a fixed transmit power p_t. The HAP's transmit beamforming can be optimized to enhance RF power transfer to different relays during the energy harvesting phase. The relay-assisted information transmission follows a two-hop half-duplex protocol. In the first hop, the HAP broadcasts signals to the relays. In the second hop, the relays forward the information to the receiver. Each RF-powered relay is equipped with a dual-mode radio structure, capable of switching between passive backscatter communications and active RF communications by amplifying and retransmitting signals using the harvested energy, as that in [7, 16]. This hybrid capability enables flexible and energy-efficient relaying. An example scenario with two relays, one operating in passive mode and the other in active mode, is illustrated in Fig. 3.1. Let $\mathbf{f}_0 \in \mathbb{C}^K$ denote the complex channel vector from the HAP (with K antennas) to the receiver, and $\mathbf{f}_n \in \mathbb{C}^K$ the channel from the HAP to the n-th relay. The scalar $g_n \in \mathbb{C}$ represents the complex channel from the relay-n to the receiver, and $z_{nm} \in \mathbb{C}$ denotes the channel between relay-n and relay-m.

3.3.1 *Time-Switching (TS) for Active Relays*

As shown in Fig. 3.1, the TS protocol divides each transmission block into distinct phases. A sub-slot of duration w (normalized to the block length) is allocated for the HAP to beamform RF energy exclusively to the active relays, enabling them to harvest energy for subsequent transmission. The remaining portion $(1 - w)$ of the block is split into two equal sub-slots, each of duration $t = (1 - w)/2$, corresponding to a two-hop half-duplex relaying protocol. In the first hop, the HAP transmits information signals using a beamforming vector $\mathbf{w}_1$. These signals are received by both the active relays and the destination. Simultaneously, the passive relays may backscatter data using the incident signal. In the second hop, the active relays collaboratively forward the amplified versions of the received signal to the destination.

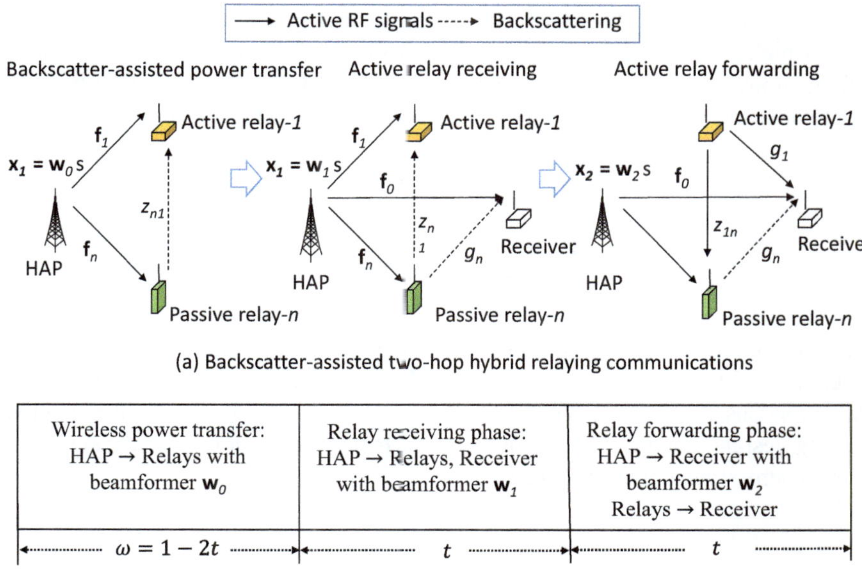

(a) Backscatter-assisted two-hop hybrid relaying communications

(b) Time-switching (TS) protocol for energy harvesting relays

Fig. 3.1 Backscatter-assisted hybrid relay communications with energy harvesting capabilities. The TS protocol divides transmission into wireless power transfer and two-hop relaying phases. During wireless power transfer, the HAP's energy beamforming $\mathbf{w}_o$ is used to power active relays. In relay communications, the HAP transmits information using different beamforming vectors $\mathbf{w}_1$ and $\mathbf{w}_2$ during both hops

Notably, the HAP continues to transmit the same information during this phase using a potentially different beamforming vector $\mathbf{w}_2$. This is motivated by the presence of a strong direct link $\mathbf{f}_0$ from the HAP to the receiver, which contributes significantly to the received signal in both hops, consistent with observations in [16]. By transmitting in both hops, the HAP enables maximal ratio combining (MRC) at the receiver, which coherently combines the directly received signal and the relayed signal to enhance reception reliability. Importantly, the optimal beamforming vectors $\mathbf{w}_1$ and $\mathbf{w}_2$ may differ in the two hops, as they serve distinct purposes. We use $\mathbf{w}_1$ to balance information delivery to relays and the destination, while use $\mathbf{w}_2$ to enhance the relay signal under updated channel conditions.

3.3.2 *Channel Enhancement via Passive Relays*

Let the binary variable $b_k \in \{0, 1\}$ indicate the operational mode of relay-k, where $b_k = 0$ corresponds to active mode and $b_k = 1$ to passive backscatter communications. Based on these mode selections, we define $\mathcal{N}_a$ and $\mathcal{N}_b$ as the sets of active and passive relays, respectively. Let $\hat{\mathbf{f}}_0 \in \mathbb{C}^K$ and $\hat{\mathbf{f}}_n \in \mathbb{C}^K$ denote the backscatter-assisted effective channels from the HAP to the receiver and to the active relay-n, respectively, which incorporate the signal components backscattered by the passive relays. Following the model in [16], these enhanced channels $\hat{\mathbf{f}}_0$ and $\hat{\mathbf{f}}_n$ account for both the direct paths and the reflected components, which can be represented as follows:

$$\hat{\mathbf{f}}_0 = \mathbf{f}_0 + \sum_{k \in \mathcal{N}} b_k g_k \Gamma_k \mathbf{f}_k, \tag{3.1}$$

$$\hat{\mathbf{f}}_n = \mathbf{f}_n + \sum_{k \in \mathcal{N}} b_k z_{kn} \Gamma_k \mathbf{f}_k, \quad \forall n \in \mathcal{N}. \tag{3.2}$$

Given a fixed set of passive relays and their reflection coefficients, the enhanced direct channel $\hat{\mathbf{f}}_0$ and the effective channels $\hat{\mathbf{f}}_n$ to the active relays can be evaluated. This allows us to decouple the passive backscattering component from the active relaying design, enabling the optimization to focus solely on the active relays and their transmission control strategies.

3.3.3 Backscatter-Aided Throughput Maximization

In the first hop, the HAP transmits the signal $\mathbf{x}_1 = \sqrt{p_t}\mathbf{w}_1 s$, where s denotes the unit-power complex information symbol. The signal is received by both the active relays and the receiver. The resulting SNR at the receiver is given $\gamma_1 = p_t|\hat{\mathbf{f}}_0^H \mathbf{w}_1|^2$, where $\hat{\mathbf{f}}_0^H$ denotes the Hermitian transpose of the enhanced direct channel $\hat{\mathbf{f}}_0$ as defined in (3.1), and the noise power is normalized to unity. In the second hop, the active relays jointly amplify and forward the received signals to the destination. Each active relay n applies an amplification (or power allocation) coefficient $x_n \in (0, 1)$, scaling its received signal before retransmission. Simultaneously, the HAP retransmits the same information symbol s to the receiver using a potentially different beamforming vector $\mathbf{w}_2$. The corresponding transmit signal from the HAP is thus $\mathbf{x}_2 = \sqrt{p_t}\mathbf{w}_2 s$, which is combined with the relay signals to enable coherent combining at the receiver. Then, the received signal r_d at the destination in the second hop is given by $r_d = \sum_{n=1}^{N} \hat{g}_n x_n r_n + \sqrt{p_t}||\hat{\mathbf{f}}_0||s + v_d$, where the sum is taken over all active relays and $\hat{g}_n \in \mathbb{C}$ denotes the enhanced channel from relay-n to the receiver, incorporating any backscatter-induced components. Assuming perfect CSI and coherent combining, the achievable SNR in the second hop is then simplified as follows:

$$\gamma_2 = \frac{\left|\sum_{n \in \mathcal{N}} x_n y_n \hat{g}_n + \sqrt{p_t}||\hat{\mathbf{f}}_0^H||\right|^2}{1 + \sum_{n \in \mathcal{N}} |x_n \hat{g}_n|^2}, \tag{3.3}$$

where we define $y_n \triangleq \sqrt{p_t}\hat{\mathbf{f}}_n^H \mathbf{w}_1$ as the received signal at the relay during the first hop. The power amplification coefficient of active relay-n is given by $x_n = \left(\frac{p_n}{1+|y_n|^2}\right)^{1/2}$, where p_n denotes its transmit power. This normalization ensures that the relay's transmit power satisfies its power constraint.

We aim to maximize the end-to-end throughput by jointly optimizing the HAP's beamforming vectors $(\mathbf{w}_0, \mathbf{w}_1)$, the time allocation t, the relays' operating modes and their reflection coefficients (b_n, θ_n). The optimization problem can be formulated as follows:

$$\max_{\mathbf{w}_0,\mathbf{w}_1,b_n,t,\theta_n} \quad t\log_2(1+\gamma_1+\gamma_2) \tag{3.4a}$$

$$s.t. \quad ||\mathbf{w}_0|| \leq 1 \text{ and } ||\mathbf{w}_1|| \leq 1, \tag{3.4b}$$

$$p_n \leq \eta(1/t-2)p_t|\hat{\mathbf{f}}_n^H\mathbf{w}_0|^2, \forall\, n \in \mathcal{N}_a, \tag{3.4c}$$

$$t \in (0, 1/2), \tag{3.4d}$$

$$b_n \in \{0, 1\}, \quad \forall\, n \in \mathcal{N}, \tag{3.4e}$$

$$\theta_n \in [0, 2\pi], \quad \forall\, n \in \mathcal{N}_b. \tag{3.4f}$$

The constraint in (3.4b) defines the feasible set of the HAP's beamforming vectors in two hops. Constraints in (3.4c) and (3.4d) model the transmit power constraints of the active relays in the second hop. Specifically, the transmit power p_n the active relay-n is limited by the amount of energy it harvested during the energy transfer phase. This energy is proportional to the duration ω of the energy harvesting slot, the HAP's transmit power p_t, and the channel condition $\hat{\mathbf{f}}_n$. The constant η denotes the energy conversion efficiency. The constraint in (3.4f) reflects the practical controllability of the complex reflection coefficient $\Gamma_n = |\Gamma_n|e^{j\theta_n}$ at each passive relay through load impedance modulation [7]. As shown in (3.1) and (3.2), the phase θ_n plays a crucial role in coherently aligning the backscattered signal with the direct or relayed components. To maximize the strength of the reflected signal, the magnitude $|\Gamma_n|$ can be fixed at its maximum allowable value, typically close to 1 for efficient backscattering. This allows the optimization to focus on phase control θ_n, simplifying the design while maintaining high performance.

3.4 Lower Bound via Monotonic Optimization

Even with a fixed radio mode selection b_n, the joint optimization of the beamforming vectors $(\mathbf{w}_0, \mathbf{w}_1)$, the passive relays' phase shifts θ_n, and time allocation remains highly challenging. This difficulty arises due to the coupled dependencies among the active relays through the beamforming design, the enhanced channels, and the end-to-end SNR. To address this complexity, we proceed in two stages. First, with the relay modes $\{b_n\}$ fixed, we derive a tractable lower bound on the optimal value of problem (3.4) by jointly optimizing the beamforming strategies, phase shifts, and time allocation. This lower bound serves as a performance benchmark and can be used to devise the optimization-driven DRL framework for solving problem (3.4) with the enhanced efficiency and reward performance.

Given the set $\mathcal{N}_b$ of passive relays and their reflection coefficients Γ_n, the enhanced channels for active RF communications can be updated according to (3.1) and (3.2), respectively. These enhanced channels incorporate both the direct links and the backscatter components, capturing the impact of passive relaying on the active communication paths. Then, we can decouple the passive backscattering design from the active relaying optimization. The throughput maximization problem reduces to the transmission control over the active relays, including the HAP's beamforming and time allocation strategies. The resulting problem is formulated as follows:

$$\max_{t,\mathbf{w}_0,\mathbf{w}_1} \quad t\log_2(1+\gamma_1+\gamma_2) \tag{3.5a}$$

$$s.t. \quad p_n \le \eta(1/t-2)p_t|\hat{\mathbf{f}}_n^H\mathbf{w}_0|^2, \quad \forall\, n\in\mathcal{N}_a, \tag{3.5b}$$

$$0<t<1/2, \ ||\mathbf{w}_0||\le 1, \text{ and } ||\mathbf{w}_1||\le 1. \tag{3.5c}$$

Although problem (3.5) is non-convex and challenging to solve optimally, a lower bound on its optimal value can still be derived and serve as a useful benchmark for evaluating the achievable throughput performance. This motivates the formulation of the following tractable optimization problem:

Proposition 3.1 *A feasible lower bound on (3.5) is given as follows:*

$$\max_{t,\mathbf{w}_0,\mathbf{w}_1} \quad t\log_2\left(1+p_t||\hat{\mathbf{f}}_0||^2+p_t\sum_{n\in\mathcal{N}_a\cup\{0\}} s_{n,1}\right) \tag{3.6a}$$

$$s.t. \quad \psi_n(1/t-2)s_{n,0}\ge s_{n,1}(1+p_t s_{n,1}), \quad \forall\, n\in\mathcal{N}_a, \tag{3.6b}$$

$$0<t<1/2, \ ||\mathbf{w}_0||\le 1, \ and \ ||\mathbf{w}_1||\le 1. \tag{3.6c}$$

where $\psi_n \triangleq \eta p_t|\hat{g}_n|^2||\hat{\mathbf{f}}_0||^2$ *and* $s_{n,i}\triangleq|\hat{\mathbf{f}}_n^H\mathbf{w}_i|^2$ *for* $i\in\{0,1\}$ *and* $n\in\{0\}\cup\mathcal{N}_a$.

The proof follows a similar approach to that in [23], where the received SNR γ_2 in the second hop is reformulated as a Rayleigh quotient. By leveraging the optimal value of this quotient, we can derive an achievable lower bound on γ_2, which facilitates tractable optimization. Additional technical details can be found in [1]. Despite this simplification, problem (3.6) remains non-convex due to the coupling between the time allocation variable t and the beamforming vector $\mathbf{w}_1$, which affects the effective channel gain. While convexity cannot be guaranteed, the problem exhibits a useful structural property, i.e., monotonicity in certain transformed variables. As shown in [24], monotonic optimization can exploit such structures to design efficient algorithms even in the absence of convexity. Specifically, by introducing a change of variables, we treat the effective SNR $\bar{\gamma}$ as a decision variable, representing the combined outcome of beamforming and relaying gains. Under this transformation, the objective function (3.6a) can be rewritten in a simpler form: $r(t,\bar{\gamma}) = t\log_2(1+\bar{\gamma})$, which is jointly increasing in t and $\bar{\gamma}$. Then, the problem can be reformulated over the feasible set defined as below:

$$\Omega \triangleq \left\{ (t, \bar{\gamma}) \;\middle|\; \begin{array}{l} \text{(3.6b) and (3.6c) hold,} \\ \bar{\gamma} \le p_t \|\hat{\mathbf{f}}_0\|^2 + p_t \sum_{n \in \mathcal{N}_a \cup \{0\}} s_{n,1}, \\ s_{n,i} \le |\hat{\mathbf{f}}_n^H \mathbf{w}_i|^2, \quad i \in \{0, 1\}, n \in \mathcal{N}_a \cup \{0\}. \end{array} \right\} \tag{3.7}$$

The transformed objective $r(t, \bar{\gamma})$ is strictly increasing in both t and $\bar{\gamma}$. This monotonicity implies that its optimum will be achieved on the boundary of Ω. This structural property enables the use of monotonic optimization techniques, such as polyblock approximation [24], to efficiently converge to the global optimum by iteratively tightening bounds on the feasible region.

3.4.1 Successive Polyblock Approximation

Monotonic optimization tackles the problem by successively approximating the feasible set Ω by using a sequence of regular polyblocks, where each polyblock is a union of finitely many box sets [24]. A box set is defined as $[\mathbf{0}, \mathbf{v}]$, where $\mathbf{v}$ is the end point of the box and also called a vertex of the polyblock. Let polyblock $P \supset \Omega$ be an approximation of the feasible set Ω. Given that the objective $r(t, \bar{\gamma})$ is strictly increasing in both t and $\bar{\gamma}$, its maximum over a polyblock is attained at one of the vertices of the polyblock P. Specifically, for a polyblock $P \supset \Omega$, the maximum of $r(t, \bar{\gamma})$ over P provides an upper bound on the optimal value of the original problem, i.e.,

$$\max_{(t,\bar{\gamma}) \in \Omega} r(t, \bar{\gamma}) \le \max_{(t,\bar{\gamma}) \in P} r(t, \bar{\gamma}) = \max_{(t,\bar{\gamma}) \in V} r(t, \bar{\gamma}), \tag{3.8}$$

where V denotes a discrete set of all vertices of the polyblock P. The first inequality in (3.8) holds due to the inclusion $\Omega \subset P$, i.e., any feasible point in Ω is also contained in P, so the maximum over P serves as an upper bound. The equality $\max_{(t,\bar{\gamma}) \in P} r(t, \bar{\gamma}) = \max_{(t,\bar{\gamma}) \in V} r(t, \bar{\gamma})$ follows from the monotonicity of the objective function $r(t, \bar{\gamma})$. As it is strictly increasing in both t and $\bar{\gamma}$, its maximum over the polyblock P will be achieved at one of the vertices in the vertex set V. Although directly maximizing $r(t, \bar{\gamma})$ over the non-convex feasible set Ω is computationally challenging, this upper bound can be efficiently computed by evaluating $r(t, \bar{\gamma})$ at each vertex $\mathbf{v} \in V$. This vertex-wise evaluation transforms a complex functional optimization into a finite search problem, enabling the design of iterative polyblock approximation algorithm that successively refine P to converge to the global optimum.

To approximate the optimal value $\max_{(t,\bar{\gamma}) \in \Omega} r(t, \bar{\gamma})$, the upper bound in (3.8) suggests an iterative strategy, i.e., we can successively construct smaller polyblocks that more tightly outer-approximate the feasible set Ω. As the polyblocks shrink toward Ω, the upper bound converges to the true optimum. To proceed, we first verify that the feasible set Ω as defined in (3.7) is a normal set, which bears the following structural property:

Proposition 3.2 *Given* $(t, \bar{\gamma}) \in \Omega$, *we have* $(t', \bar{\gamma}') \in \Omega$ *for* $(t', \bar{\gamma}') \preceq (t, \bar{\gamma})$.

The detail proof can be found in [1]. This structural property is essential for monotonic optimization, as it ensures that the Pareto boundary of Ω can be approached from above via a sequence of polyblocks. The successive generation of new polyblocks follows a similar procedure as that in [25]. In particular, if we have some $\mathbf{z} \in P$ and $\mathbf{z} \notin \Omega$, for any $\mathbf{z}' \succeq \mathbf{z}$, we can ensure that $\mathbf{z}' \notin \Omega$ by the normal structural property in Proposition 3.2. This suggests that we can construct a tighter outer approximation P' from the current polyblock, i.e., by cutting off a box set from P:

$$P' = P \setminus \{\mathbf{z}' \in P : \mathbf{z}' \succeq \mathbf{z}\}. \tag{3.9}$$

This ensures $P \supset P' \supset \Omega$. The initial polyblock P_0 can be simply chosen as the smallest box set that covers the whole feasible set Ω. Given the time allocation constraints in (3.7), we have $t \leq 1/2$. Similarly, let $\bar{\gamma}_{\max}$ denote a valid upper bound on the achievable SNR $\bar{\gamma}$ under the system's power and channel constraints. Hence, the vertex of the polyblock P_0 is given by $(1/2, \bar{\gamma}_{\max})$. From (3.7), an analytically tractable bound can be derived by relaxing coupling constraints. A simple yet effective choice is given as follows:

$$\bar{\gamma}_{\max} = 2p_t||\hat{\mathbf{f}}_0||^2 + p_t \sum_{n \in \mathcal{N}_a} ||\hat{\mathbf{f}}_n||^2.$$

In the k-th iteration of the polyblock approximation algorithm, we first compute an upper bound r_k^U on the optimal objective value by evaluating the objective function $r(t, \bar{\gamma})$ at the maximizing vertex of the current polyblock P_k. Specifically, $r_k^U = \max_{\mathbf{z} \in V_k} r(\mathbf{z})$, where V_k denotes the set of vertices of P_k. As such, we can easily determine the optimal vertex $\mathbf{z}_k = \arg\max_{\mathbf{v} \in V_k} r(\mathbf{v})$ and the corresponding upper bound $r_k^U = r(\mathbf{z}_k)$. Due to the monotonicity of $r(t, \bar{\gamma})$, this maximum r_k^U over P_k provides a valid upper bound on the optimal value of the original problem. A lower bound r_k^L on the optimal throughput is obtained by projecting the maximizing vertex of the current polyblock onto the feasible set Ω, as detailed in the next subsection. This projection yields a feasible solution whose objective value serves as a achievable lower bound. During the successive polyblock approximation process, the lower bound r_k^L and the upper bound r_k^U are iteratively tightened. As the polyblock P_k shrinks and better approximates the boundary of Ω, the gap between r_k^L and r_k^U monotonically decreases and the algorithm terminates when this gap $r_k^U - r_k^L$ falls below a pre-specified tolerance.

3.4.2 *Solving SDPs for Feasibility Check*

The primary computational challenge in monotonic optimization lies in obtaining a tight lower bound r_k^L at each iteration, which requires finding a feasible point in Ω that is as close as possible to the current maximizing vertex. While general projection onto Ω may be difficult, we show that for our problem structure this lower bound can be efficiently computed by solving a sequence of semidefinite programs (SDPs), leveraging convex reformulations of the underlying beamforming and power allocation constraints. Specifically, let $\mathbf{z}_k = (t_k, \bar{\gamma}_k)$ denote the vertex of the current polyblock P_k that maximizes the objective $r(\mathbf{z}_k)$. If $\mathbf{z}_k \in \Omega$, then it is not only feasible but also globally optimal (since it achieves the current upper bound). In this case, we conclude that $r_k^L = r_k^U$, and the algorithm can terminate immediately with a certificate of global optimality. Otherwise, if $\mathbf{z}_k \notin \Omega$, we solve an SDP feasibility or projection problem to find a feasible pair $(t, \bar{\gamma}) \in \Omega$ such that $r(t, \bar{\gamma})$ is maximized, thereby yielding a non-trivial lower bound $r_k^L < r_k^U$.

Specifically, when the maximizing vertex $\mathbf{z}_k = (t_k, \bar{\gamma}_k)$ lies outside the feasible set Ω, we project it onto the boundary of Ω along the ray from the origin to $\mathbf{z}_k$ by introducing a scaling factor $\lambda \in (0, 1)$. The projection aims to find the largest scaling factor λ_k such that the scaled point remains feasible, i.e., $\lambda_k = \arg\max\{\lambda \in (0, 1] : (\lambda t_k, \lambda \bar{\gamma}_k) \in \Omega\}$. The corresponding projection point is then given by $\mathbf{o}_k \triangleq \lambda_k \mathbf{z}_k$, and the lower bound is updated as $r_k^L = r(\mathbf{o}_k)$. Due to the normal structural property of Ω, it is easy to see that $\lambda \mathbf{z}_k \in \Omega$ for $\lambda \in (0, \lambda_k]$ and $\lambda \mathbf{z}_k \notin \Omega$ for $\lambda \in (\lambda_k, 1]$. This implies a bisection method for an efficient computation of the maximum scaling factor λ_k. At each bisection step, for a given $\lambda \in (0, 1)$, we check whether $\lambda \mathbf{z}_k \in \Omega$. If a feasible solution exists, we increase λ in next bisection step, otherwise we decrease it. The process continues until λ_k is determined within a desired accuracy. By the structural property of Ω, the feasibility test $(\lambda t_k, \lambda \bar{\gamma}_k) \in \Omega$ given any λ is equivalent to solve the following subproblem:

$$\max_{\mathbf{w}_0, \mathbf{w}_1} \quad p_t \|\hat{\mathbf{f}}_0\|^2 + p_t \sum_{n \in \mathcal{N}_a \cup \{0\}} s_{n,1} \tag{3.10a}$$

$$\text{s.t.} \quad \begin{bmatrix} \psi_n \left(\frac{1}{\lambda t_k} - 2\right) s_{n,0} - s_{n,1}, & \sqrt{p_t} s_{n,1} \\ \sqrt{p_t} s_{n,1}, & 1 \end{bmatrix} \succeq 0, \tag{3.10b}$$

$$s_{n,i} \le |\hat{\mathbf{f}}_n^H \mathbf{w}_i|^2 \text{ and } \|\mathbf{w}_i\| \le 1, \tag{3.10c}$$

$$i \in \{0, 1\} \text{ and } n \in \mathcal{N}_a \cup \{0\}. \tag{3.10d}$$

The constraint in (3.10b) is equivalent to (3.6b) and is expressed as a linear matrix inequality (LMI), which is convex in the matrix variables. By applying semidefinite relaxation (SDR), i.e., relaxing any non-convex rank-one constraints on the beamforming matrices, we can reformulate problem (3.10) as a convex semidefinite SDP. This convex relaxation enables efficient global solution via standard SDP solvers, such as SeDuMi or MOSEK, with polynomial-time complexity.

Let m_k denote the optimum value of the relaxed SDP (3.10). Then, the projection point $(\lambda t_k, \lambda \bar{\gamma}_k) \in \Omega$ is considered feasible with respect to the original constraint set Ω if $m_k \ge \lambda \bar{\gamma}_k$. This condition indicates that there exists a feasible beamforming and power allocation strategy capable of achieving the scaled SNR $\lambda \bar{\gamma}_k$ under the time allocation λt_k. Given the projection point $\mathbf{o}_k = \lambda_k \mathbf{z}_k$, the updated polyblock P_{k+1} is constructed by cutting off the outer region dominated by $\mathbf{o}_k$ from the current polyblock P_k. This procedure follows the standard polyblock reduction rule as defined in (3.9). In particular, we construct a separating cone defined as $P_k^c \triangleq \{\mathbf{z}|\mathbf{z} \succeq \mathbf{o}_k\}$ such that $P_k^c \cap \Omega = \emptyset$, which contains all vertex points that dominate the projection point $\mathbf{o}_k$. A new polyblock P_{k+1} is then constructed by cutting off $\Delta_k \triangleq P_k^c \cap P_k = \{\mathbf{z} \in P_k|\mathbf{z} \succeq \mathbf{o}_k\}$ from the current polyblock P_k, i.e., $P_{k+1} = P_k \setminus \Delta_k$, which preserves the polyblock structure while eliminating infeasible points that dominate $\mathbf{o}_k$. The vertex set of P_{k+1} is updated accordingly to ensure that the outer approximation of Ω remains tight.

3.5 Optimization-Driven Hierarchical DRL

The most straightforward DRL solution to the mixed-integer non-convex optimization problem (3.4) is to reformulate it into an MDP and deploy a single centralized agent at the HAP, which observes the system state and jointly decides the HAP's beamforming and the relaying strategies simultaneously. The goal is to maximize the long-term expected sum rate through end-to-end policy learning. At the beginning of the t-th decision epoch, the system state observed by the agent is denoted as $\mathbf{s}_t = (\mathbf{c}_t, \mathbf{e}_t)$, where $\mathbf{c}_t$ represents the CSI and $\mathbf{e}_t = [e_{1,t}, e_{2,t}, \ldots, e_{N,t}]^T$ captures the energy status of the N relays. The channel state include the direct channel $\hat{\mathbf{f}}_0$ from the HAP to the destination and each relay's channel conditions $(\hat{\mathbf{f}}_n, \hat{g}_n)$, for $n \in \mathcal{N}$. Thus, we denote $\mathbf{c}_t = \{\hat{\mathbf{f}}_0, (\hat{\mathbf{f}}_n, \hat{g}_n)_{n\in\mathcal{N}}\}$. The energy state $e_{n,t}$ of relay-n evolves according to the energy harvesting and consumption dynamics, depending on the initial energy $e_{n,t-1}$ and the RF energy $h_n \triangleq \eta(1-2t)p_t|\hat{\mathbf{f}}_n^H \mathbf{w}_0|^2$ harvested from the HAP's beamforming during the first phase, as define in (3.4c). Each relay's energy state also depends on its power consumption p_n for relay communications, related to the relay's operating mode. For active mode ($b_n = 1$), the power consumption p_n is strongly related to the data rate, while it remains a small fixed value (e.g., circuitry power) for passive relays ($b_n = 0$). We assume that both $\mathbf{c}_t$ and $\mathbf{e}_t$ are perfectly measurable at the start of each decision epoch, enabling full state observation for the agent. The action taken by the agent at time t is given by $\mathbf{a}_t = (t, \mathbf{w}_0, \mathbf{w}_1, \mathbf{q}_t)$, which consists of the relaying strategy $\mathbf{q}_t$, the HAP's the time allocation t, and beamforming strategies $(\mathbf{w}_0, \mathbf{w}_1)$ for energy and information transfer in two hops. The relaying strategy $\mathbf{q}_t = \{b_n, \theta_n\}_{n\in\mathcal{N}}$ includes each relay's binary mode selection $b_n \in \{0, 1\}$ (active forwarding vs. passive backscattering) and the operating parameter θ_n (e.g., reflection coefficient or amplification gain). Except the binary mode selection, all other decision variables are continuous. This results in a hybrid action space, combining discrete b_n and continuous $(t, \mathbf{w}_0, \mathbf{w}_1, \theta_n)$ components. The feasibility of action $\mathbf{a}_t$ is constrained by the physical limits in problem (3.4), including transmit power budgets, energy causality, and SNR requirements. The immediate reward at epoch t is defined as the achievable throughput in that decision cycle, i.e., $v_t(\mathbf{s}_t, \mathbf{a}_t) = t\log_2(1 + \gamma_1 + \gamma_2)$, where γ_1 and γ_2 denote the effective SNRs from the direct link and the two-hop relayed links, respectively, evaluated based on the current action $\mathbf{a}_t$ and system state $\mathbf{s}_t$. By aligning the reward with the objective of problem (3.4), the cumulative discounted return approximates the long-term average throughput. Thus, maximizing this return enables the agent to learn a policy that approximates the optimal solution to the original optimization problem (3.4).

3.5.1 Hierarchical Integration of DQN and DDPG

The original optimization problem (3.4) is a mixed-integer non-convex program, involving both discrete and continuous decision variables. Specifically, the binary variable $b_n \in \{0, 1\}$ determines the operating mode of relay-n, i.e., $b_n = 1$ for active forwarding and $b_n = 0$ for passive backscattering. The continuous variables include the HAP's time allocation strategy and beamforming vectors $(\mathbf{w}_0, \mathbf{w}_1)$ for energy transfer and information forwarding, respectively. The passive relays' reflection coefficients $\boldsymbol{\theta} = [\theta_1, \theta_2, \ldots, \theta_N]^T$ are also continuous and high-dimensional. This results in a hybrid action space that poses significant challenges for conventional DRL algorithms. Directly applying DQN requires enumerating all mode combinations, leading to an exponential explosion in the action space. Conversely, standard policy-gradient methods cannot handle discrete decisions natively.

To address this challenge, we propose a hierarchical DDPG (H-DDPG) framework that decouples the discrete and continuous components of the decision process. The action vector is split into two parts:

$$\mathbf{a}_t = (\mathbf{b}_t, \mathbf{a}_t^c),$$

where $\mathbf{b}_t = [b_1, b_2, \ldots, b_N]^T$ is the discrete mode selection vector, determined by an outer-loop DQN agent, while $\mathbf{a}_t^c = (t, \mathbf{w}_0, \mathbf{w}_1, \boldsymbol{\theta})$ is the continuous control vector, optimized by an inner-loop DDPG agent. The outer-loop DQN first selects a relay activation pattern $\mathbf{b}_t$ based on the current system state $\mathbf{s}_t$. Given the outer-loop relay mode selection $\mathbf{b}_t$, the inner-loop DDPG solves the resulting continuous subproblem in (3.5) with fixed relay modes, by learning an optimal policy for $\mathbf{a}_t^c$. The long-term reward (e.g., achievable rate) obtained by the inner-loop agent is then used to compute the Q-value for the selected $\mathbf{b}_t$, enabling the outer-loop DQN agent to improve its mode selection policy continuously. A key insight of the H-DDPG framework is that the converged performance of the inner-loop DDPG serves as the effective Q-value for that discrete action in the outer loop. This creates a tight coupling between the two learning layers, where the inner loop evaluates candidate relay mode selections and the outer loop explores and exploits them over time. Such a hierarchical learning facilitates a structured decomposition of the original problem, i.e., the combinatorial optimization of relay operating modes is decoupled from the continuous resource allocation involving beamforming, time scheduling, and reflection design, effectively avoiding the exponential growth of the action space. The main advantage lies in the dimensionality reduction, i.e., the outer-loop DQN only explores over binary mode selections, while the inner-loop DDPG only refines continuous actions. As a result, this hierarchical framework achieves significantly improved sample efficiency, training stability, and convergence speed, making it well-suited for large-scale wireless networks with high-dimensional hybrid action spaces.

3.5.2 Optimization-Driven Inner-Loop DDPG

Conventional DDPG often suffers from slow convergence and training instability, primarily due to the random initialization of its actor and critic networks. In particular, the target Q-value y_t, generated by the target Q-network, may deviate significantly from the true optimal value, especially in the early stages of training. This results in high-variance learning signals, leading to inefficient exploration and prolonged convergence. To address this issue, we enhance the inner-loop DDPG by incorporating model-based optimization into the target value estimation, which provides a better-informed target value for DNN training.

Specifically, at the t-th decision epoch, the actor network produces a candidate action $\mathbf{a}_t^c = (t, \mathbf{w}_0, \mathbf{w}_1, \boldsymbol{\theta})$. Instead of relying solely on the target Q-network for y_t, we leverage the optimization framework developed in Sect. 3.4 to compute an approximate solution $\mathbf{a}_t^o = (t', \mathbf{w}_0', \mathbf{w}_1', \boldsymbol{\theta}')$ to problem (3.5). From this solution, we derive an optimization-driven target value y_t', which serves as a lower bound estimate of the achievable performance under the current state and mode selection. This model-based estimate y_t' is expected to be more accurate and physically meaningful than the randomly initialized or poorly trained target y_t from the target Q-network, particularly during initial learning phases when policy knowledge is limited. The integration of model-based optimization offers two key advantages. Firstly, by providing a better-informed target y_t', the agent receives higher-quality feedback early in training, accelerating policy improvement and reducing sample complexity. Moreover, the derivation of y_t' is independent of the online Q-network and relies only on the problem structure and current state. This decoupling eliminates feedback loops between the online and target networks, making y_t' inherently more stable than the DNN-generated y_t. As a result, the learning process experiences less fluctuations, leading to smoother convergence and reduced variance in reward performance.

Figure 3.2 illustrates the main components of the proposed optimization-driven H-DDPG framework. The architecture consists of two interconnected loops: the outer-loop DQN and the inner-loop optimization-driven DDPG. This hierarchical structure enables efficient learning in hybrid action spaces by decoupling combinatorial decisions from continuous control. At the t-th decision epoch, given the current system state $\mathbf{s}_t$, the outer-loop DQN selects a relay mode configuration $\mathbf{b}_t$. Based on $\mathbf{b}_t$, the channel conditions are updated to reflect the resulting relay configuration including both active forwarding and backscattering links. These updated channel conditions, along with system constraints and transmission performance, form the input state for the inner-loop DDPG, which generates the continuous action $\mathbf{a}_t^c$ and

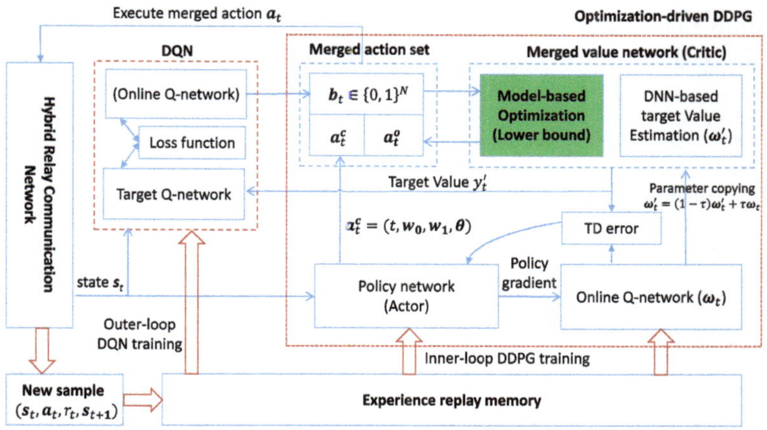

Fig. 3.2 The optimization-driven hierarchical DDPG (H-DDPG) framework comprises outer-loop DQN and inner-loop DDPG, enhanced by a model-based optimization module. The outer loop DQN selects relay modes $\mathbf{b}_t$, while the inner loop DDPG generates continuous actions $\mathbf{a}_t^c$ using updated channel conditions. Model-based optimization provides approximate solutions $\mathbf{a}_t^o$ and target values y_t' to enhance learning efficiency, stabilize training, and accelerate convergence by combining neural network outputs with optimization-driven targets

the value estimate, respectively. DDPG's actor network generates a candidate action $\mathbf{a}_t^c$. The critic network estimates its corresponding Q-value, producing a DNN-based target y_t for training updates. To enhance learning efficiency, we incorporate the model-based approximate solution $\mathbf{a}_t^c = (t', \mathbf{w}_0', \mathbf{w}_1', \boldsymbol{\theta}')$ derived from the optimization framework in Sect. 3.4, which yields an optimization-driven target value y_t'. A hybrid action selection rule is then applied based on a comparison between the values of y_t and y_t'. Considering the simplest selection rule, if $y_t' > y_t$, indicating that the model-based solution outperforms the current policy, we replace the neural network output with the optimized action $\mathbf{a}_t = (\mathbf{b}_t, \mathbf{a}_t^o)$, and use y_t' as the target value for DDPG training. Otherwise, when $y_t' < y_t$ (e.g., in later training stages with a well-trained actor and stable performance), we retain the actor's output and proceed with the standard DDPG updates. This adaptive mechanism ensures that the agent benefits from high-quality initial guidance while preserving the flexibility of end-to-end learning. When the inner-loop DDPG becomes stabilizes, its converged performance under a fixed $\mathbf{b}_t$ provides a reliable estimate of the expected return for that mode configuration. This value is fed back to the outer-loop DQN as a more accurate Q-value estimate, effectively accelerating mode selection learning. If the DDPG agent consistently underperforms the model-based lower bound $y_t < y_t'$, it indicates poor policy exploration or insufficient training, prompting corrective measures (e.g., learning rate adjustment or replay buffer prioritization). The complete procedure is summarized in Algorithm 1.

The implementation of the proposed optimization-driven H-DDPG aligns with the system architecture illustrated in Fig. 3.3. A hierarchical learning agent at the HAP makes sequential decisions based on the perceived transmission performance,

Algorithm 1 Optimization-driven H-DDPG for Hybrid Relay Networks

Initialize Q-networks of DQN and DDPG
Initialize the replay buffer for DQN and DDPG
DQN loop
  DQN agent selects relays' operating modes $\mathbf{b}_t$
  DDPG loop
    DDPG agent generates continuous action $\mathbf{a}_t^c$
    Execute action $\mathbf{a}_t^c \leftarrow \mathbf{a}_t^c + \mathcal{N}_0$
    Evaluate the reward r_t of the action $\mathbf{a}_t^c$
    Update y_t' and $\mathbf{a}_t^o$ by adaptive optimization or from the cached results
    If $y_t' > y_t$
      Accept optimization-driven target $y_t \leftarrow y_t'$
      Adopt the merged action $\mathbf{a}_t \leftarrow (\mathbf{b}_t, \mathbf{a}_t^o)$
    END If
    Buffer transition sample in replay memory
    Sample a minibatch from replay memory
    Update critic and actor networks
  END DDPG loop
  Sample a minibatch from DQN's replay memory
  Update DQN network and its target network
END DQN loop

as well as the channel and energy states of the relays. We adopt a block-fading channel model, where the channel conditions remain constant within each transmission period but may vary across epochs. At the beginning of each decision epoch, the HAP collects up-to-date channel information and all relays' energy status. Leveraging this information, the agent jointly optimizes the relay mode selection, beamforming vectors, and reflection coefficients. The resulting strategies are then disseminated to the relays via a downlink control channel. Throughput estimation is performed based on feedback signals (e.g., ACK/NACK or measured SNR) reported by the receiver. To enhance learning efficiency, the HAP also hosts an optimization module to evaluate an approximate solution of the continuous subproblem under the current mode configuration. Under partial or imperfect system information, the model-based optimization yields both a candidate action and a lower bound on the achievable throughput performance, providing informed guidance to the DDPG agent during training.

3.5.3 Discussion on Practical Implementation

3.5.3.1 Decentralized Implementation

The proposed H-DDPG framework relies on centralized training at the HAP, which raises scalability concerns as the number of relays grows. In the standard model-free H-DDPG, the continuous action space of the inner-loop DDPG agent includes beamforming and reflection variables for all relays, leading to exponentially increasing action dimension as the number of relays increases. We can also adopt a decentralized multi-agent DRL approach [26], where each relay acts as an autonomous agent making decisions based on local observations. Instead of learning all continuous actions from scratch, an optimization module is integrated in the inner loop to efficiently computes an approximate joint action vectors, including beamforming and phase shifts of all relays, by solving a one-shot SDP relaxation of the original problem. This allows the system to bypass the curse of dimensionality in policy search, as the optimization scales polynomially rather than exponentially with the number of relays. Furthermore, the framework supports a hybrid execution strategy, i.e., only a subset of relays (e.g., those in dynamic or uncertain environments) are actively controlled by the DDPG agent, while the remaining relays adopt model-based optimization-driven actions directly. This selective learning approach reduces the effective action space and accelerates training without sacrificing performance.

3.5.3.2 Light-Weight Q-Value Approximation

The optimization module depicted in Fig. 3.2 provides an estimate of the performance lower bound and generates informative action targets to guide the learning process. Specifically, it computes an approximate solution under the current relay mode configuration, enabling efficient estimation of the inner-loop control variables, such as beamforming vectors and reflection coefficients, even when system information (e.g., CSI or energy states) is incomplete or imperfectly known. Although the full joint optimization problem is generally non-convex and high-dimensional, rendering it intractable for real-time solution, fixing a subset of variables often allows the remaining subproblem to be relaxed into a convex form that can be solved efficiently. This conditional tractability motivates the hierarchical decomposition of control variables into outer and inner loops. Importantly, the choice of variable splitting is not arbitrary. It should reflect the underlying physical coupling and functional dependencies among control actions. For instance, relay selection modes fundamentally alter the network topology and thus strongly influence optimal beamforming and signal forwarding strategies. This justifies the assignment of mode selection to the outer loop and continuous resource allocation to the inner loop. However, such decomposition must be designed on a case-by-case basis, considering both problem structure and computational complexity.

A key practical principle for decomposing control variables in a hierarchical structure is that the inner-loop optimization need not be highly accurate, but must be computationally lightweight and fast. High-precision solutions often require complex algorithms and iterative procedures, which introduce significant latency per decision epoch and hinder real-time learning convergence. Therefore, while monotonic optimization can provide globally optimal solutions for problem (3.6), we instead pursue a lightweight approximation that sacrifices marginal accuracy for substantial gains in speed. Specifically, we consider a simplified variant of problem (3.6) by fixing the transmission time variable t and optimizing only the remaining beamforming vectors $(\mathbf{w}_0, \mathbf{w}_1)$ and phase shifts $\boldsymbol{\theta}$ in the optimization module of Fig. 3.2. With a fixed t, the resulting subproblem reduces to a SDP in the form of (3.10), which can be solved efficiently using standard interior-point methods [27]. This simplification has a dramatic impact on runtime, whereas monotonic optimization requires multiple SDP solves across iterations. Such a fixed-time approximation performs only one time per decision epoch, significantly reducing the run time for the inner-loop optimization. As shown in Simulation results of Sect. 3.6, this improves the learning efficiency significantly without notable performance degradation. Remarkably, the model-simplified H-DDPG still outperforms the model-free counterpart and achieves performance very close to that of the H-DDPG framework driven by monotonic optimization. These results demonstrate that our learning framework is robust to variations in optimization fidelity. Even coarse or approximate guidance improves sample efficiency and convergence. When full system information is unavailable practically, we can still obtain substantial benefits by incorporating partial domain knowledge via simplified models into the inner-loop optimization.

3.5.3.3 Adaptive Inner-Loop Optimization

In addition to lightweight Q-value approximation, adaptive execution of the inner-loop optimization offers another effective way for improving learning efficiency. Although the SDP in (3.10) is computationally tractable compared to the original non-convex problem (3.4), its solution still incurs non-negligible latency. This can significantly prolong the overall runtime of the optimization-driven H-DDPG algorithm, especially when the optimization module is invoked in every iteration. To mitigate this bottleneck, we introduce two enhancements within Algorithm 1.

- On-demand adaptive optimization: we can employ an adaptive scheme that triggers the model-based optimization module on demand. The inner-loop optimization provides an approximate solution conditioned on the time allocation variable t. When the time allocation t slightly changes, the optimal beamforming and phase shift strategies typically varies smoothly. Therefore, frequent and computationally intensive re-optimizations in consecutive iterations are unnecessary, as the solution from a similar configuration remains a valid and high-quality approximation. We therefore adopt an adaptive scheme that executes the optimization module only when t undergoes a significant change. This makes line 9 of Algorithm 1 conditional rather than mandatory, reducing redundant computation without sacrificing solution quality.
- Parallel execution with caching: To further decouple optimization from learning, we enable asynchronous operation through a caching mechanism. Specifically, the optimization module runs as a background subroutine, e.g., line 9 of Algorithm 1 can be executed in parallel with the main learning loop on a separate computational thread or process. Upon a triggering event, it solves the SDP and stores the resulting action tuple $(\mathbf{w}_0, \mathbf{w}_1, \boldsymbol{\theta})$ in a shared cache. The inner-loop learning agent can access the cached optimization results and dynamically decide whether to adopt, or ignore them during online decision-making. This mechanism decouples the learning process from the optimization latency, ensuring that each learning

epoch proceeds without blocking on the completion of the inner-loop optimization module. Instead, it proceeds immediately by either adopting a cached optimization solution when the outer-loop variables remain sufficiently stable, or resorting to action exploration or policy-based inference if no valid approximation is available. Given the typically slow evolution of DNN-parameterized policy networks during training, cached optimization solutions remain reasonably accurate across consecutive decision epochs making this approach both efficient and robust.

3.6 Numerical Evaluation

In the experiment, we consider a two-hop relay-assisted wireless network consisting of a 3-antenna HAP and 5 hybrid relays, as illustrated in Fig. 3.3. The hybrid relays can harvest RF energy from the HAP's signal beamforming for self-powering and reflect or retransmit information to the destination. The geometric layout of all nodes, including the HAP, relays, and destination, is specified in Fig. 3.3. The RF energy harvesting efficiency at each relay is set to 0.6, and the HAP's transmit power varies

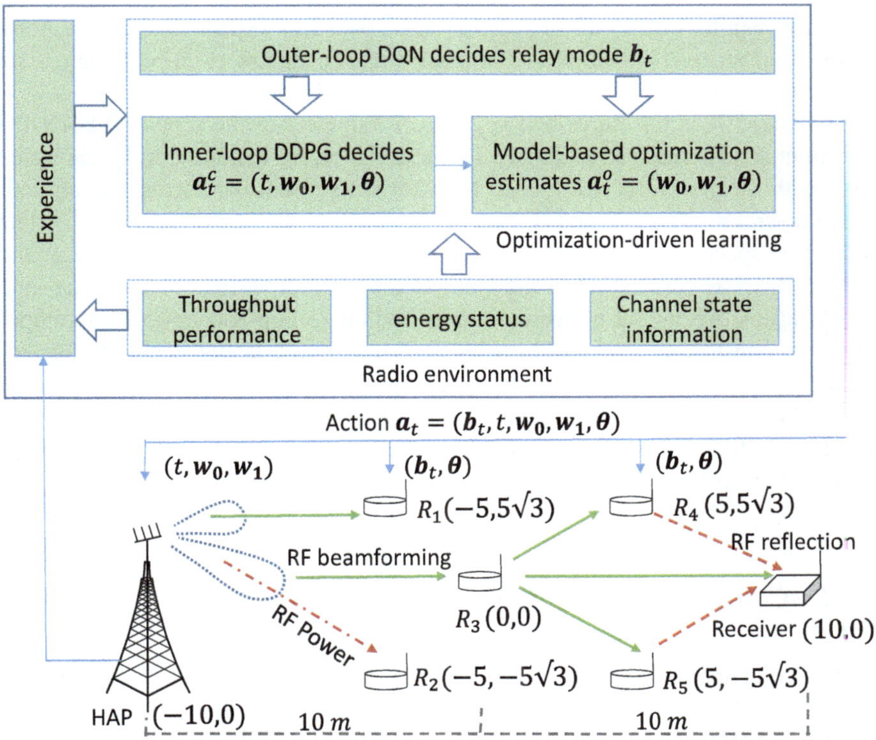

Fig. 3.3 Experimental configuration of a hybrid relay network and the application of H-DDPG for optimizing relay mode selection, joint beamforming, and time allocation

Table 3.1 Network structure and hyperparameters in H-DDPG

	Component	Structure	Hyperparameter	Value
Outer-loop (DQN)	DNN	`fc(N_STATES, 64)`	Batch size	32
		`relu`	Learning rate	0.01
		`fc(64, 64)`	Reward discount	0.9
		`relu`	Memory capacity	150
		`fc(64, N_ACTIONS)`	Exploration rate	0.1
Inner-loop (DDPG)	Critic	`fc(STATES_DIM, 64)`	Batch size	32
		`fc(ACTIONS_DIM, 64)`	Learning rate (critic)	0.002
		`relu, fc(64, 1)`	Learning rate (actor)	0.001
	Actor	`fc(STATES_DIM, 64)`	Reward discount	0.9
		`fc(64, ACTIONS_DIM)`	Memory capacity	150
		`tanh`	Replacement factor	0.005

between -10 dBm to 10 dBm. We adopt a log-distance path loss model with a reference path loss of 25 dB at 1 meter and a path loss exponent of 2. The noise power at the receiver is -80 dBm. For passive relays, the reflection coefficient is set to 0.5, modeling moderate signal attenuation during backscattering.

The network architectures and key hyperparameters of the proposed H-DDPG algorithm are summarized in Table 3.1, where `fc(m,n)` denotes a fully connected layer with the size of $m \times n$. `relu` and `tanh` represent the corresponding activation layers. `N_STATES` and `N_ACTIONS` denote the dimensionality of the state and action spaces for the outer-loop DQN agent, respectively, `STATES_DIM` and `ACTIONS_DIM` correspond to the state and action dimensions of the inner-loop DDPG agent. All neural networks in both the outer-loop and inner-loop components are constructed exclusively using fully connected layers, without convolutional or recurrent structures. This design choice simplifies hardware implementation and enables straightforward computation of the model's computational complexity. Specifically, the total number of adders and multipliers required for inference can be directly derived from the sizes of the fully connected layers.

3.6.1 Convergence of Optimization-Driven H-DDPG

We first compare the reward performance of the optimization-driven H-DDPG algorithm with that of the conventional model-free H-DDPG, as illustrated in Fig. 3.4a. The solid lines depict the average reward over 80 training episodes, while the shaded regions represent the fluctuation range of the rewards across independent runs. The optimization-driven H-DDPG significantly enhances learning efficiency by leverag-

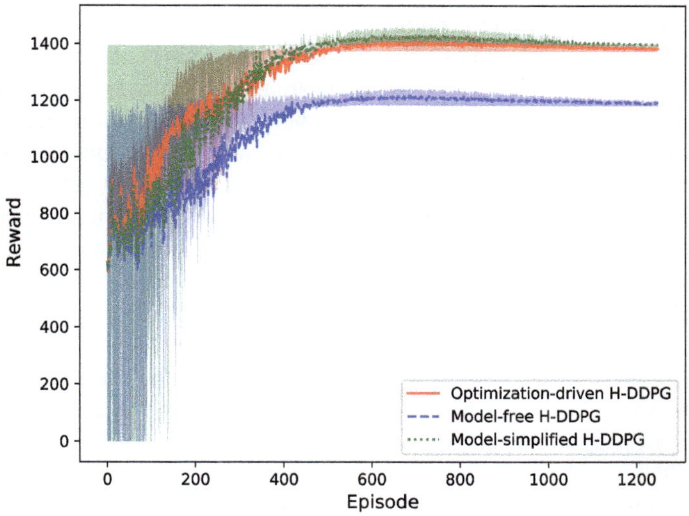

(a) Achievable reward

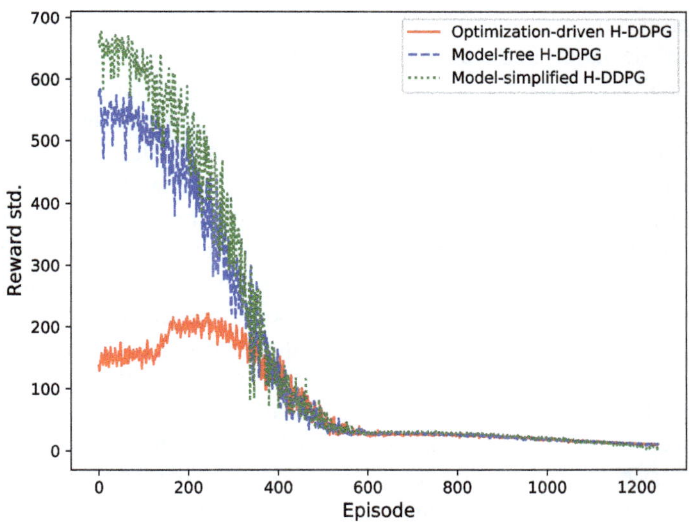

(b) Training stability

Fig. 3.4 Comparison of convergence behavior, achieved reward, and training stability among different H-DDPG algorithm variants. The optimization-driven H-DDPG significantly outperforms model-free H-DDPG, achieving over 20% higher reward with lower variance. It leverages monotonic optimization for informed Q-value estimates, enhancing learning efficiency and stability. The model-simplified variant achieves similar performance but exhibits higher variance during the initial learning phase

ing monotonic optimization to compute a feasible solution to problem (3.4), which provides a tighter lower bound on the optimal value. This bound serves as a more informed estimate of the target Q-value during training, effectively guiding the policy updates with model-based insights. Building upon the discussion in Sect. 3.5.3, we further introduce a model-simplified H-DDPG variant. In this approach, the inner-loop DDPG agent first generates a candidate time allocation strategy t, which is then held fixed while the optimization module computes the optimal beamforming vectors $(\mathbf{w}_0, \mathbf{w}_1)$ and phase shifts $\boldsymbol{\theta}$. By partially optimizing the control subproblem, the optimization module efficiently produces a performance lower bound with significantly reduced computational complexity, enabling faster integration into the learning loop.

Although the rewards of all three H-DDPG variants exhibit fluctuations during training due to exploratory actions, they eventually converge to stable values as the agents learn from accumulated experience. A key observation is that the optimization-driven H-DDPG achieves a significantly higher converged reward, over 20% greater than that of the model-free H-DDPG, demonstrating its effectiveness of integrating model-based optimization into the model-free learning framework. Moreover, the shaded region (representing reward variability across runs) for the optimization-driven H-DDPG is noticeably narrower than that of its model-free counterpart, indicating improved training stability. This is further validated in Fig. 3.4b, which plots the standard deviation (std.) of the episode rewards. The optimization-driven H-DDPG exhibits consistently lower reward variance, particularly in the early and middle stages of training. An interesting trend is observed in the variance curve of the optimization-driven H-DDPG. It initially increases slightly before dropping sharply to its minimum. This behavior can be explained as follows: in the early phase, the DDPG agent frequently adopts action suggestions from the optimization module, which provides consistent and high-quality target Q-values derived from monotonic optimization. These targets are less sensitive to environmental noise and exploration randomness, resulting in low variance. As learning progresses, the agent begins to explore more aggressively around these informed baselines, temporarily increasing its reward and also the variance. Eventually, the policy converges to a high-performance regime with reduced fluctuations. When comparing the two model-based H-DDPG variants, the model-simplified H-DDPG achieves reward performance very close to that of the monotonic optimization driven version. However, their learning dynamics differ significantly. The model-simplified variant exhibits higher reward variance, especially during the initial learning phase. As shown in Fig. 3.4b, this increased instability stems from the fact that its optimization module relies on the time allocation strategy t generated by the inner-loop DDPG agent. During early training, t varies widely due to random exploration, leading to inconsistent optimization targets and amplified reward fluctuations. This highlights a trade-off between computational efficiency and training stability by integrating partial model information and model-based optimization into model-free learning frameworks.

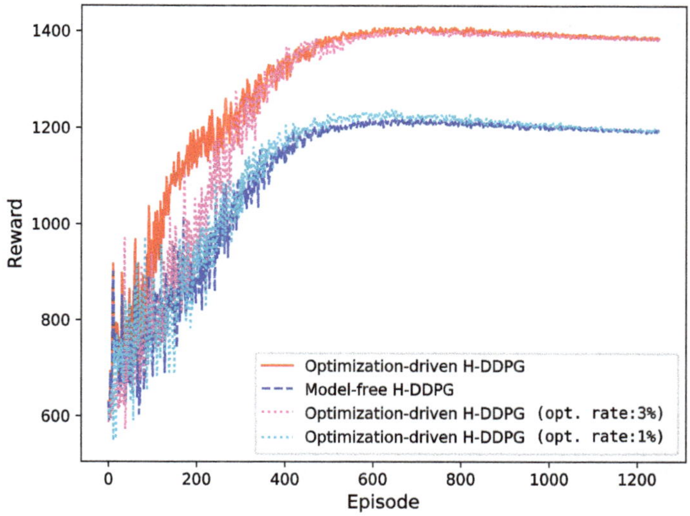

Fig. 3.5 Reward in Algorithm 1 with adaptive optimization and caching. Adaptive optimization reduces computational burden by executing optimization periodically with rate λ. At $\lambda = 3\%$, the optimization-driven H-DDPG achieves near-full performance with drastically reduced runtime. Even at $\lambda = 1\%$, it remains competitive while offering better convergence, balancing reward quality and computational efficiency

3.6.2 Adaptive Inner-Loop Optimization

Table 3.2 compares the average execution time per learning episode across different algorithm variants. As SDP in (3.10) involves high computational complexity, the optimization-driven H-DDPG incurs significantly higher runtime when the optimization module is invoked in every episode. In contrast, the model-simplified H-DDPG optimizes only a subset of variables given a fixed time allocation, achieving a 40 times improvement in execution efficiency over the monotonic optimization driven H-DDPG. Compared to the model-free H-DDPG, the model-simplified variant requires approximately three times longer per episode, due to the runtime cost of the inner-loop optimization. However, this additional computational cost is justified by the substantially higher converged reward, as demonstrated in Fig. 3.4a, highlighting the trade-off between performance and efficiency. To mitigate the computational burden of the optimization-driven H-DDPG, we implement an adaptive optimization and caching mechanism in Algorithm 1, as discussed in Sect. 3.5.3. Instead of solving the optimization problem in every decision epoch, the optimization module is executed only periodically. We define the optimization rate $\lambda \in [0, 1]$ as the ratio of the number of optimization calls to the total number of learning episodes. A higher λ indicates more frequent invocation of the optimization module. When $\lambda = 0$, the algorithm reduces to the model-free H-DDPG. Figure 3.5 evaluates the

Table 3.2 Average run time in each episode (in milliseconds)

Number of hybrid relays	5	4	3	2	1
Model-free H-DDPG	1.25	1.24	1.22	1.20	1.19
Model-simplified H-DDPG	5.08	4.05	3.92	3.67	3.25
Optimization-driven H-DDPG	227.64	214.15	199.39	185.32	168.34
Optimization-driven H-DDPG (opt. rate: 3%)	6.68	6.37	6.03	5.50	5.10
Optimization-driven H-DDPG (opt. rate: 1%)	2.22	2.12	2.04	1.84	1.71

reward performance under different λ values. At $\lambda = 3\%$, the optimization-driven H-DDPG achieves nearly the same converged reward as the full-frequency version, while drastically reducing average runtime (see Table 3.2). Even at $\lambda = 1\%$, both the reward and run time performance of the optimization-driven H-DDPG remain competitive with the model-free H-DDPG, while offering slightly better convergence performance. These results confirm that a sparse but informed and adaptive use of model-based optimization can deliver near-optimal learning performance with dramatically reduced computational overhead, effectively balancing the trade-off between reward quality, training stability, and execution efficiency.

3.6.3 Throughput via Direct Link and Hybrid Relays

In this section, we evaluate the optimal relay-assisted throughput as a function of the HAP's transmit power p_t. For benchmarking, we also present the maximum achievable throughput when only the direct link (DL) is utilized. To model varying DL channel quality, we introduce an additional path loss factor L_e, representing attenuation due to obstacles or blockages. A higher L_e (e.g., 40 dB) corresponds to a severely degraded DL channel. As shown in Fig. 3.6a, the end-to-end throughput increases with p_t for all configurations, which is expected due to improved signal strength. Notably, the performance gain from relay-assisted transmission becomes more pronounced as the DL channel weakens. For instance, when $L_e = 40$ dB and $p_t = -10$ dBm, the hybrid relay-assisted transmission achieves a 345% higher throughput

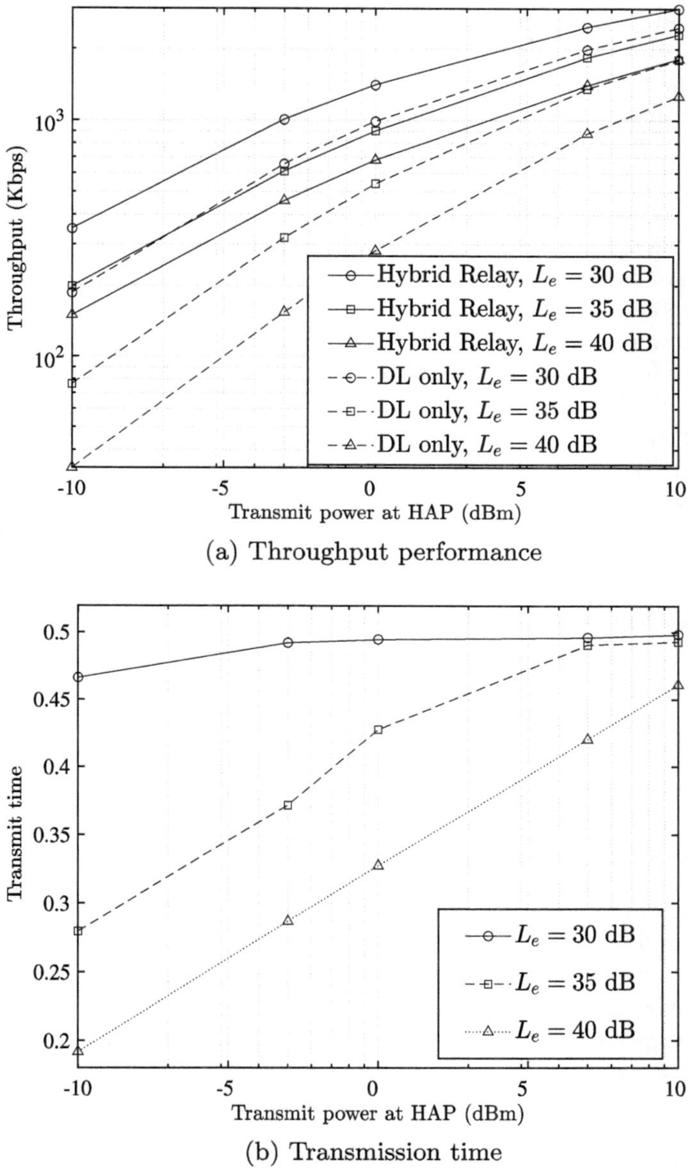

Fig. 3.6 Hybrid relays contribute more significantly to the end-to-end throughput under weak DL channel conditions and limited transmit power at the HAP, where their cooperative forwarding compensates for poor channel quality and enhances overall system throughput

compared to the DL-only case. This gain decreases to 43.48% when p_t increases to $p_t = 10$ dBm, indicating that relay-assisted transmission is most beneficial under low transmit power and poor DL channel conditions. The diminishing relative gain at high transmit power p_t arises because a strong DL channel can already achieve near-optimal throughput, reducing the relays' marginal contribution to the overall throughput. Conversely, when the DL is weak or the HAP's transmit power is limited, the hybrid relays play a critical role in enhancing both throughput and energy efficiency. Figure 3.6b illustrates the optimal time allocation between WPT and information transmission. As p_t increases, the relays' energy harvesting time decreases, allowing more time to be allocated for data forwarding. Moreover, under a fixed transmit power p_t, the DL channel quality significantly influences the relays' optimal energy harvesting time. When the DL channel is degraded with a larger L_e, e.g., $L_e = 40$ dB, the HAP will allocate more WPT time to the relays. This enables the hybrid relays to harvest more energy and enhance their contribution to the overall throughput. This adaptive scheduling highlights the intelligent resource allocation enabled by the proposed framework.

3.6.4 Comparison of Different Beamforming Schemes

We compare the optimal throughput achieved by the optimization-driven H-DDPG algorithm against several baseline beamforming schemes under varying DL channel conditions, as illustrated in Fig. 3.7a. To evaluate robustness across different DL qualities, we vary the additional path loss L_e from 30 to 40 dB, where higher values correspond to increasingly degraded DL channels. The following baselines are considered in the performance comparison:

- Random scheme: For the simplest implementation, the HAP selects its beamforming vector uniformly at random.
- Max-DL scheme: The HAP's beamformer is simply aligned with the DL channel to maximize direct transmission rate.
- Max-Energy scheme: The same beamformer is used in both WPT and information transmission phases, i.e., $\mathbf{w}_0 = \mathbf{w}_1$, simplifying implementation while prioritizing relays' energy harvesting.

All algorithms employ the same hierarchical framework for outer-loop relay selection as the optimization-driven H-DDPG, ensuring fairness in comparison. The key difference lies in the inner-loop beamforming design strategy, which directly affects computational complexity during learning. In terms of complexity, the optimization-driven H-DDPG algorithm solves the SDP (3.10) iteratively within each learning episode. Based on the interior-point method analysis in [27], its complexity is $O(2^{1.5}(KN)^{2.5}(K + N))$, where K is the number of HAP's antennas and N is the number of active relays. The Max-Energy scheme requires solving a simplified version of the SDP (3.10) under the constraint $\mathbf{w}_0 = \mathbf{w}_1$, reducing the number of variables

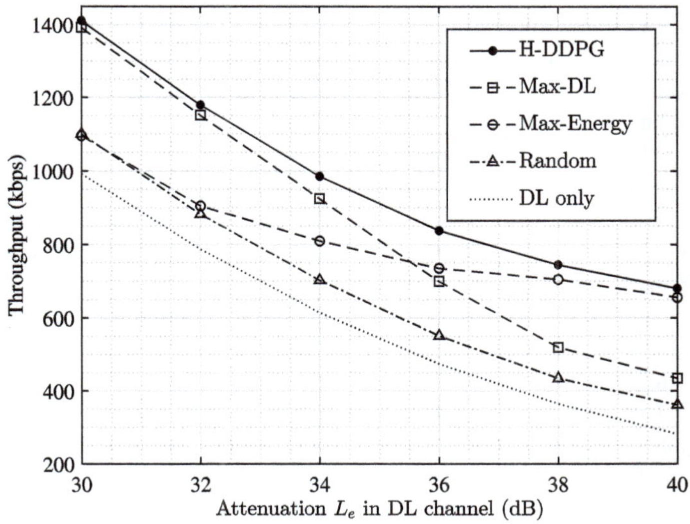

(a) Throughput comparison

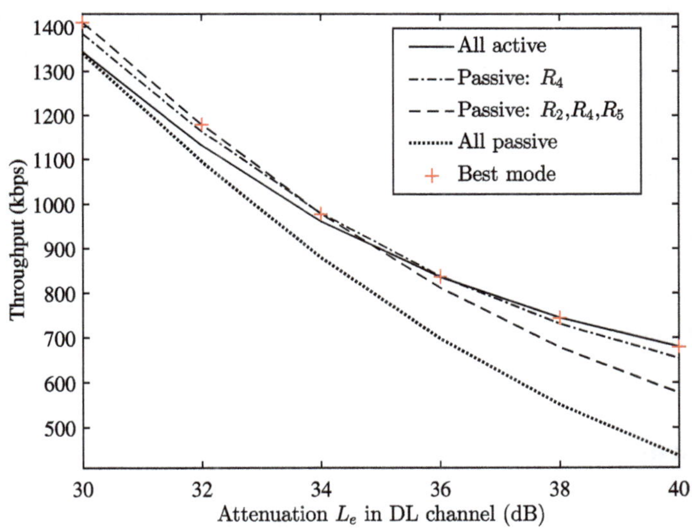

(b) Relay mode selection

Fig. 3.7 The optimal throughput and the relays' operational mode selection depend critically on the DL channel quality. As L_e varies, the system dynamically adapts beamforming and hybrid relay strategies to maximize end-to-end throughput performance

and yielding a complexity of $O((KN)^{2.5}(K+N))$. The random and the Max-DL schemes involve only closed-form or heuristic decisions, resulting in constant-time complexity $O(1)$. While the optimization-driven H-DDPG incurs higher per-episode computation, it enables more flexible and performance-optimal beamforming design. The Max-Energy scheme offers a compromise between complexity and performance. The Random and Max-DL schemes are computationally lightweight and easy to implement, but they fail to jointly optimize energy and information transfer, leading to significant throughput loss, especially under weak DL channel conditions.

As shown in Fig. 3.7a, the optimization-driven H-DDPG achieves the highest throughput, serving as the performance benchmark. The Max-DL scheme performs comparably to the optimal scheme when the DL channel is strong (i.e., low attenuation L_e). However, its performance degrades significantly as L_e increases, indicating its inability to adapt to poor direct links effectively. In contrast, the Max-Energy scheme exhibits an opposite trend, i.e., its relative performance improves as L_e increases, and the gap to the optimal scheme narrows under severely degraded DL conditions. This is because, under weak DL channels, prioritizing energy transfer to relays becomes more beneficial. These contrasting behaviors highlight the adaptability of the H-DDPG algorithm, which dynamically balances beamforming between enhancing DL channel and powering relays based on real-time channel conditions. We further compare these schemes to the DL-only baseline. Notably, even the Random beamforming scheme in the relay-assisted network outperforms the DL-only case. This demonstrates that hybrid relays inherently provide cooperative diversity gain, improving end-to-end reliability and throughput even without intelligent beamforming, thereby validating the value of relay-assisted cooperative transmissions in resource-limited wireless networks. Figure 3.7b illustrates how the relays' optimal mode selection evolves with varying DL channel quality. As L_e increases and the DL channel weakens, more relays become active in AF modes, while fewer remain in passive backscattering mode. This dynamic adaptation enables the hybrid relay network to compensate for the degraded DL by enhancing cooperative transmission. The results demonstrate that the proposed framework achieves optimal throughput by allowing each relay to autonomously and optimally configure its operating mode based on channel conditions.

3.7 Discussions and Conclusions

In this chapter, we have proposed an optimization-driven H-DDPG algorithm for throughput maximization in hybrid relay networks incorporating both active and passive relays. The proposed framework integrates DQN and model-based optimization techniques into a hierarchical learning architecture, enhancing the model-free DDPG algorithm with structured guidance. A key innovation lies in the optimization module, which computes a performance lower bound using partial system information. This lower bound serves as a more informed target estimation for Q-value, enabling the online Q-network to achieve faster convergence and improved stability during training. As a case study, the framework is applied to a multi-relay-assisted hybrid radio network, where joint beamforming, relay selection, and time allocation are optimized to maximize end-to-end throughput. Simulation results demonstrate that the proposed algorithm significantly outperforms the model-free H-DDPG in terms of learning efficiency and robustness.

References

1. S. Gong, Y. Zou, J. Xu, D.T. Hoang, B. Lyu, D. Niyato, Optimization-driven hierarchical learning framework for wireless powered backscatter-aided relay communications. IEEE Trans. Wirel. Commun. **21**(2), 1378–1391 (2022)
2. X. Lu, D. Niyato, H. Jiang, D.I. Kim, Y. Xiao, Z. Han, Ambient backscatter assisted wireless powered communications. IEEE Wirel. Commun. **25**(2), 170–177 (2018)
3. R. Long, H. Guo, G. Yang, Y.-C. Liang, R. Zhang, Symbiotic radio: a new communication paradigm for passive internet of things. IEEE Internet Things J. **7**(2), 1350–1363 (2019)
4. G. Yang, D. Yuan, Y.-C. Liang, R. Zhang, V.C.M. Leung, Optimal resource allocation in full-duplex ambient backscatter communication networks for wireless-powered IoT. IEEE Internet Things J. **6**(2), 2612–2625 (2019)
5. J. Guo, X. Zhou, S. Durrani, Wireless power transfer via mmWave power beacons with directional beamforming. IEEE Wirel. Commun. Lett. **8**(1), 17–20 (2019)
6. D.T. Hoang, D. Niyato, P. Wang, D.I. Kim, Z. Han, Ambient backscatter: a new approach to improve network performance for RF-powered cognitive radio networks. IEEE Trans. Commun. **65**(9), 3659–3674 (2017)
7. S. Gong, J. Xu, D. Niyato, X. Huang, Z. Han, Backscatter-aided cooperative relay communications in wireless-powered hybrid radio networks. IEEE Netw. **33**, 234–241 (2019)
8. G. Yang, Q. Zhang, Y.C. Liang, Cooperative ambient backscatter communications for green internet-of-things. IEEE Internet Things J. **5**(2), 1116–1130 (2018)
9. X. Jia, X. Zhou, Performance characterization of relaying using backscatter devices. IEEE Open J. Commun. Soc. **1**, 819–834 (2020)
10. B. Lyu, D.T. Hoang, Optimal time scheduling in relay assisted batteryless IoT networks. IEEE Wirel. Commun. Lett. **9**(5), 706–710 (2020)
11. B. Lyu, D.T. Hoang, Z. Yang, Backscatter then forward: a relaying scheme for batteryless IoT networks. IEEE Wirel. Commun. Lett. **9**(4), 562–566 (2020)
12. X. Lu, D. Niyato, H. Jiang, E. Hossain, P. Wang, Ambient backscatter-assisted wireless-powered relaying. IEEE Trans. Green Commun. Netw. **3**(4), 1087–1105 (2019)
13. D. Li, Two birds with one stone: exploiting decode-and-forward relaying for opportunistic ambient backscattering. IEEE Trans. Commun. **68**(3), 1405–1416 (2020)
14. B. Lyu, D.T. Hoang, Z. Yang, User cooperation in wireless-powered backscatter communication networks. IEEE Wirel. Commun. Lett. **5**(3), 2015–2024 (2018)
15. S. Gong, L. Gao, J. Xu, Y. Guo, D.T. Hoang, D. Niyato, Exploiting backscatter-aided relay communications with hybrid access model in device-to-device networks. IEEE Trans. Cogn. Commun. Netw. **5**(4), 835–848 (2019)
16. S. Gong, Y. Zou, D.T. Hoang, J. Xu, W. Cheng, D. Niyato, Capitalizing backscatter-aided hybrid relay communications with wireless energy harvesting. IEEE Internet Things J. **7**(9), 8709–8721 (2020)
17. N. Van Huynh, D.T. Hoang, D.N. Nguyen, E. Dutkiewicz, D. Niyato, P. Wang, Optimal and low-complexity dynamic spectrum access for RF-powered ambient backscatter system with online reinforcement learning. IEEE Trans. Commun. **67**(8), 5736–5752 (2019)
18. N. Van Huynh, D.N. Nguyen, D.T. Hoang, E. Dutkiewicz, Jam me if you can: defeating jammer with deep dueling neural network architecture and ambient backscattering augmented communications. IEEE J. Sel. Area. Commun. **37**(11), 2603–2620 (2019)
19. N. Van Huynh, D. N. Nguyen, D. Thai Hoang, E. Dutkiewicz, M. Mueck, Ambient backscatter: a novel method to defend jamming attacks for wireless networks. IEEE Wirel. Commun. Lett. **9**(2), 175–178 (2020)

20. T.T. Anh, N.C. Luong, D. Niyato, Y. Liang, D.I. Kim, Deep reinforcement learning for time scheduling in RF-powered backscatter cognitive radio networks, in *Proceedings of IEEE WCNC*, Apr 2019, pp. 1–7
21. F. Jameel, W.U. Khan, S.T. Shah, T. Ristaniemi, Towards intelligent IoT networks: reinforcement learning for reliable backscatter communications, in *Proceedings of IEEE GLOBECOM Workshops*, Dec 2019, pp. 1–6
22. Y. Xie, Z. Xu, Y. Zhong, J. Xu, S. Gong, Y. Wang, Backscatter-assisted computation offloading for energy harvesting IoT devices via policy-based deep reinforcement learning, in *Proceedings of IEEE/CIC ICCC Workshops*, Aug 2019, pp. 65–70
23. X. Luo, J. Xu, Y. Zou, S. Gong, L. Gao, D. Niyato, Collaborative relay beamforming with direct links in wireless powered communications, in *Proceedings of IEEE WCNC*, Apr 2019, pp. 1–6
24. Y.J.A. Zhang, L. Qian, J. Huang, Monotonic optimization in communication and networking systems. Found. Trends Netw. **7**(1), 1–75 (2013)
25. J. Xu, Y. Zou, S. Gong, L. Gao, D. Niyato, W. Cheng, Robust transmissions in wireless-powered multi-relay networks with chance interference constraints. IEEE Trans. Commun. **67**(2), 973–987 (2019)
26. A.A. Khan, R. Adve, Centralized and distributed deep reinforcement learning methods for downlink sum-rate optimization. IEEE Trans. Wirel. Commun. **19**(12), 8410–8426 (2020)
27. Z.-Q. Luo, W.-K. Ma, A.-C. So, Y. Ye, S. Zhang, Semidefinite relaxation of quadratic optimization problems. IEEE Signal Process. Mag. **27**(3), 20–34 (2010)

Chapter 4
Hierarchical DRL for IRS-Assisted AoI Minimization

4.1 Introduction

In this chapter, our case study focuses on minimizing AoI in an IRS-assisted and wirelessly powered sensor network coordinated by a multi-antenna AP. Each sensor node is capable of generating sensing data and reporting the most recent information to the AP using energy harvested from the AP's RF beamforming. Our objective is to minimize the average AoI by jointly optimizing sensor nodes' scheduling and transmission control strategies. The use of IRS not only enhances channel conditions for data transmission but also improves the efficiency for wireless power transfer, by an joint control of the AP's RF beamforming and the IRS's phase shift matrix. The following analysis and detailed algorithm design have been presented in our earlier work [1]. This chapter refines and reorganizes the content from [1] to ensure consistency, clarity, and coherence within the book's scope.

Considering the dynamic and unpredictable nature of sensors' data arrivals, we propose a hierarchical DRL framework for AoI minimization in energy-constrained sensor networks. This framework operates in two steps. First, an outer-loop DRL agent makes high-level scheduling decisions by determining which sensor users should be granted access to the communication channel at each time slot. This scheduling decision is based on dynamic network states, such as current AoI, channel conditions, and energy levels, enabling adaptive prioritization of different sensor nodes to ensure their timely status updates. Second, an inner-loop optimization module is invoked to handle low-level resource adaptation efficiently, either optimizing uplink data transmission for data offloading or downlink wireless energy transfer for energy replenishment, according to the high-level scheduling decisions. To enhance computational efficiency and ensure real-time applicability, we also introduce a simple and effective approximation method to reduce the runtime overhead of the inner-loop optimization significantly. Such a hierarchical learning framework integrates the efficiency of model-based optimization for real-time transmission control and the flexibility of model-free DRL for long-term scheduling. It is expected to ensure a

S. Gong et al., *Optimization-Driven Deep Reinforcement Learning for Wireless Networks*, Wireless Networks, https://doi.org/10.1007/978-3-032-22997-7_4

balance between information freshness and sustainable network operation, resulting in a practical and efficient solution for AoI-aware wireless networks, particularly for energy-constrained IoT scenarios with energy harvesting capabilities.

4.1.1 Motivations and Contributions

With the rapid advancement of future IoT technologies, a growing number of emerging applications, such as autonomous driving, interactive gaming, and virtual reality, rely critically on the timely transmission and processing of sensory data collected from IoT devices in the physical world [2]. Ensuring information freshness in these time-sensitive applications necessitates a paradigm shift from traditional latency-centric metrics to a more comprehensive performance measure. As such, AoI is defined as the elapsed time since the generation of the most recently received status update at the destination [3], thereby capturing both the delay and the timeliness of information from the perspective of the receiver. In wireless networks, AoI is influenced by two primary components, i.e., the waiting or scheduling delay before transmission and the in-the-air transmission delay. The former is largely determined by the multi-user scheduling policy and access contention, while the latter depends on network capacity, channel conditions, and physical-layer factors such as mutual interference and channel fading. To minimize transmission delay and consequently AoI, it is essential to exploit favorable channel conditions by dynamically adapting transmission control parameters, including transmit power, sub-channel allocation, and beamforming strategies. Prior work has investigated AoI-aware scheduling and transmission control using queueing theory [4], modeling the system under assumptions of specific stochastic processes for data arrivals, service times, and request patterns. However, deriving closed-form expressions for AoI performance is often mathematically tedious and limited to simplified scenarios, for example, with known probability distributions for the sensor nodes' data arrivals and service time. These analytical approaches usually struggle to generalize to more realistic and dynamic environments characterized by random user mobility, intermittent connectivity, and energy constraints.

In more complex and heterogeneous wireless networks, involving mobile users and energy-limited IoT devices, it remains highly challenging to design optimal or near-optimal multi-user scheduling policies that effectively minimize the overall AoI. The interplay between resource allocation, energy management, and time-varying channel conditions introduces a high-dimensional and non-convex optimization problem that is difficult to solve using conventional analytical optimization methods. This complexity motivates the need for adaptive and learning-based solutions capable of handling uncertainty and dynamics in real time. In energy-constrained wireless networks, minimizing AoI critically depends on the joint control of each node's energy supply and demand, particularly in wireless powered communication networks (WPCNs), where sensor nodes rely on harvested energy for data transmissions. When a sensor node is scheduled more frequently to maintain its low

AoI, it consumes more energy and demands more wireless energy transfer (WET) to sustain its operation. However, when more channel time and power are used to supply energy to certain node, this will reduce transmission opportunities and energy delivery to other nodes, potentially degrading their information freshness. This interplay arises a critical AoI-energy trade-off in energy-constrained IoT systems [5]. The joint optimization of sensor nodes' scheduling and transmission control is inherently complicated and typically formulated as a dynamic program due to the spatio-temporal couplings among sensor nodes. These dependencies stem from shared spectrum resources, time-varying channel conditions, and random data arrivals. With limited energy availability, scheduling decisions should be energy-aware to carefully balance the urgency of status updates against the risk of energy depletion. This challenge is further exacerbated by partial observability, as channel states and data arrival processes are often unknown or unpredictable. This implies that the scheduling decisions should be made and adapted based only on historical AoI and local observations. Traditional model-based optimization methods become impractical under such uncertainty and complexity.

To address these limitations, model-free DRL methods have emerged as a flexible alternative for AoI minimization, capable of reformulating the problem into MDP without requiring full system dynamics [5–7]. In a DRL framework, the agent's action typically includes both scheduling decisions and transmission control parameters, while the system state should consider AoI, energy levels, and channel conditions. In IRS-assisted wireless networks with a large-size IRS, the state and action spaces grow significantly in dimensionality, leading to unstable training, poor sample efficiency, and slow convergence with traditional DRL methods. This scalability issue highlights the need for more efficient and structured learning frameworks that can handle complexity while preserving flexibility and adaptability. Therefore, there is a compelling motivation to design a computationally efficient and scalable learning framework, by leveraging domain knowledge and model-based optimization to reduce action space and accelerate convergence.

In this case study, we aim to minimize the average AoI in a wireless-powered sensor network comprising a multi-antenna AP, a wall-mounted IRS, and multiple single-antenna sensor nodes that periodically sample status updates from distinct physical processes. The IRS is deployed to intelligently reconfigure the wireless channel conditions, thereby reducing both transmission delay and energy consumption. It can assist either the downlink WET from the AP to the sensor nodes or the uplink information transmission from the nodes to the AP [8], depending on the high-level AP's scheduling decision. The joint optimization of user scheduling and IRS-assisted beamforming poses a significant challenge, as it involves a high-dimensional mixed-integer problem with both continuous beamforming and power allocation strategies, as well as the discrete scheduling decisions. We propose a two-step hierarchical DRL framework that significantly improves learning efficiency and scalability compared to conventional model-free DRL approaches. The core idea is to use a model-free DRL algorithm such as DQN or PPO to adapt the scheduling strategy in the outer loop, while an inner-loop optimization module concurrently optimizes the beamforming strategy. The outer-loop DRL will select a subset of sensor nodes for uplink data transmission in each time slot, based on current AoI, energy levels, and channel conditions. Given the scheduling decisions from the outer loop, the model-based optimization in the inner loop performs beamforming optimization at the AP and phase shift configuration at the IRS. It either optimizes the AP's downlink energy transfer to all nodes or adapts individual node's uplink information transmission. The inner-loop optimization leverages approximation or iterative algorithms, e.g., successive convex approximation (SCA) and AO to efficiently compute near-optimal beamforming solutions, while the outer-loop DRL adapts the scheduling policy that accounts for long-term AoI and energy dynamics. By decoupling combinatorial scheduling from beamforming optimization, the proposed hierarchical learning framework reduces the dimensionality of the learning space and enables faster convergence. It not only enhances computational efficiency but also improves stability and interpretability, making it a practical solution for AoI-aware resource management in energy-constrained and IRS-assisted sensor networks.

4.1.2 Related Works

4.1.2.1 DRL for AoI Minimization

DRL has recently emerged as a powerful tool for minimizing AoI in dynamic wireless networks, enabling adaptive scheduling and resource control in response to time-varying traffic demands and channel conditions, without requiring prior knowledge about the underlying stochastic processes. For instance, the authors in [6] proposed DQN to optimize a freshness-aware scheduling scheme, where the DQN agent continuously learns and updates its scheduling policy to maximize long-term information freshness. The agent prioritizes users with rapidly increasing AoI, effectively reducing stale information across the network. The work in [7] focused on a multi-user status update system where a single sensor monitors a physical process and selectively transmits status updates to multiple receivers over time-varying channels. Without prior knowledge of channel statistics, the DRL agent at the sensor learns an optimal transmission policy by leveraging only the instantaneous ACK/NACK feedback from the receivers. The transmission policy decides both when to transmit and to which destined receiver to send the next information update. Such a feedback driven approach enables real-time adaptation under uncertain channel conditions. The work in [9] investigated a multi-access scenario in which a BS aims to maximize the timely collection of information from multiple wireless users, each subject to a strict transmission deadline. To balance timeliness and resource consumption, PPO was adopted to learn the scheduling policy from historical traffic patterns, demonstrating improved performance in AoI-sensitive scenarios. A different scheduling scheme was explored in [10], where multiple sensor nodes' simultaneous transmissions are allowed to enhance spectral efficiency. To minimize AoI, such a multi-user scheduling policy is optimized by using DDQN, which decouples action selection and evaluation using two separate DNNs. This reduces overestimation bias in the Q-value approximation and leads to more stable and accurate scheduling decisions for AoI minimization.

Furthermore, the model-free DRL offers significant flexibility in handling complex optimization problems under dynamic resource constraints. It can learn optimal policies directly from interactions with the environment, enabling robust adaptation to time-varying traffic, channel conditions, and energy availability, making it particularly well-suited for managing the AoI-energy trade-off in energy-constrained IoT systems [11]. For instance, AoI-energy trade-off was examined in [12], highlighting the potential for energy savings without significantly compromising information freshness. DQN was applied in [12] to control content update actions in the caching node, dynamically deciding whether to refresh cached data based on AoI and energy status. The authors in [5] considered an edge caching node that collects status updates from sensor nodes and serves freshness-sensitive user requests. At each time step, the caching node schedules a subset of sensors to report their latest measurements. To balance timeliness and energy efficiency, a variant of DQN was employed to optimize the scheduling policy, aiming to minimize a long-term average cost defined as a weighted sum of the average AoI and the total energy consumption of the sensor nodes. A similar composite cost function combining AoI and energy was adopted in [13], where the decision is learned dynamically to request a fresh update or serve stale data. These studies discussed above collectively demonstrate the flexibility and effectiveness of model-free DRL in addressing diverse aspects of AoI-aware network scenarios. In contrast to prior works that focus primarily on scheduling or caching decisions, the case study of this chapter considers a more comprehensive control problem that jointly optimizes dynamic scheduling and beamforming strategies for AoI minimization in a wireless-powered IoT network. The coupling of discrete scheduling and high-dimensional continuous beamforming variables leads to slow convergence and unstable training by using the model-free DRL methods. To address this, we decouple the joint optimization problem and propose a hierarchical learning framework that separates high-level scheduling from low-level beamforming control. Such a hierarchical decomposition can effectively reduce the dimensionality of the action space, thereby significantly enhancing learning efficiency, stability, and scalability.

4.1.2.2 RF-Powered Scheduling for AoI Minimization

WPT and energy harvesting have emerged as promising solutions to sustain the massive deployment of low-cost, battery-limited sensor nodes in future IoT systems. However, energy harvesting is inherently unreliable and intermittent, as the amount of harvested energy depends heavily on time-varying channel conditions, environmental factors, and the availability of transmission opportunities for WPT. This stochastic nature introduces significant challenges for the sensor nodes' control of energy consumption and information freshness, necessitating the trade-off study between energy conservation and AoI minimization. To address this challenge, the authors in [14] investigated AoI minimization in a cognitive radio network with dynamic availability of both energy and spectrum resources. Using a dynamic programming approach, they derived the optimal scheduling policy with a threshold-type structure, i.e., a sensor should be scheduled only when its AoI exceeds a certain level depending on its current energy state and channel conditions. Further theoretical analysis in [15] showed that the optimal update policy exhibits a similar threshold behavior, i.e., each sensor should transmit a status update only if its AoI exceeds a threshold that is a function of its available energy. This enforces a lazy transmission when energy is scarce and aggressive updates when energy is abundant. Considering AoI minimization under long-term energy constraints, the authors in [16] proposed a Lyapunov optimization framework to design a dynamic control algorithm that stabilizes energy queues while minimizing AoI. While effective, such model-based optimization is often computationally demanding and requires system knowledge such as energy arrival distributions or channel state information, which may not be available in practical deployments. To overcome these limitations, the authors in [17] reformulated the AoI minimization problem into MDP and proposed an online Q-learning algorithm to learn the optimal transmission policies. Such a model-free learning method gradually converges to a policy that balances AoI reduction and energy sustainability, without assuming prior knowledge of energy arrival patterns.

There is a growing trend of employing DRL to tackle AoI minimization problems in RF-powered wireless networks. The authors in [18] studied an RF-powered wireless network in which devices harvest energy from a dedicated BS during the downlink WPT and subsequently use the harvested energy to transmit status updates in the uplink. The authors employed DQN to learn an optimal scheduling strategy to minimize the long-term AoI that can balance energy supply for future updates against the urgency of current data transmissions. The authors in [19] investigated a spectrum-sharing cognitive radio system, where secondary users opportunistically access the channel to send status updates, sustained by RF energy harvesting. Both DQN and Dueling DDQN were applied in [19] to adapt sensing and information update policies. Focusing on energy harvesting ad hoc networks, the authors of [20] aimed at minimizing AoI by jointly adapting the scheduling and power allocation strategies using the A2C algorithm. The results show that A2C achieves comparable AoI performance to near-optimal solutions while significantly reducing runtime, making it suitable for real-time implementation in decentralized ad hoc networks. Integrating UAV's mobility with WPT, the authors in [21] considered a UAV-assisted and RF-powered wireless network, where an UAV acts as a mobile BS to simultaneously transfer RF energy and collect data from GUs. Model-free DQN was employed to jointly optimize the UAV's trajectory, user scheduling, and energy harvesting durations for minimizing the average AoI of all users. By leveraging the UAV's mobility to enhance both WPT and data transmission, this approach demonstrates significant gains in information freshness and shows enhanced robust against the time-varying channel conditions due to the UAV's movement.

4.1.2.3 IRS-Assisted AoI Minimization

The IRS's reconfigurability offers a powerful solution to enhance wireless communications by intelligently shaping the propagation conditions. By adjusting the phase shifts of its passive reflecting elements, an IRS can be used to improve channel quality, increase link capacity, and reduce transmission delay [8]. Despite the potential of IRS technology, only a limited number of studies have explored the application of IRS specifically for minimizing AoI in wireless networks. For instance, the authors in [22] investigated an IRS-assisted MEC system, aiming to minimize the end-to-end latency of task execution by optimizing the IRS's passive beamforming strategy. While not directly minimizing AoI, this work demonstrated that IRS's channel reconfigurability can contribute to timeliness by reducing communication latency. The authors in [23] focused on uplink transmissions under strict delay constraints, demonstrating that IRS's beamforming can effectively reduce users' transmit power while still meeting delay requirements. More directly related to AoI, the authors in [24] employed IRS to improve AoI performance via joint optimization of user scheduling and passive beamforming. A model-free DRL method was used to adapt the combinatorial scheduling decisions based on historical system states, while the IRS's passive beamforming was optimized via SDP. Although effective, this method decouples the two optimization tasks and relies on computationally intensive convex solvers for beamforming design. Similarly, the authors in [25] utilized a UAV-mounted IRS to relay information from GUs to a BS, requiring a joint optimization of UAV's altitude, transmission scheduling, and IRS's beamforming to minimize AoI. The UAV's mobility adds another degree of freedom and thus increases the complexity for real-time control.

In contrast to these prior works, focusing primarily on uplink improvements, our case study in this chapter explores performance gains in both uplink and downlink of an IRS-assisted wireless network. Specifically, the IRS's passive beamforming is not only used to enhance uplink information transmission but also to improve and balance the efficiency of downlink WPT to all users. Such a bidirectional exploitation seamlessly integrates IRS-assisted WET and information transmission within a unified optimization framework, achieving superior AoI performance while maintaining energy sustainability.

4.2 System Model and Problem Formulation

Our case study focuses on an IRS-assisted wireless sensor network deployed in smart city environments to support real-time information sensing and data-driven decision making [1]. The system comprises a multi-antenna AP equipped with M antennas, an IRS composed of N reconfigurable passive elements, and a set $\mathcal{K} \triangleq \{1, 2, \ldots, K\}$ of K single-antenna IoT devices. This architecture can be straightforwardly extended to scenarios involving multiple APs or multiple IRSs. The IoT devices act as wireless sensor nodes, for example, deployed on digital advertising panels or roadside lampposts, to monitor environmental parameters such as air quality and traffic conditions. Each node generates small-sized status updates that must be delivered to the AP in a timely manner for continuous monitoring and situational awareness. For cost-effective deployment, these sensors are not equipped with fixed energy sources. Instead, they operate using RF energy harvesting from the AP's downlink transmissions, following a wireless-powered communication paradigm similar to [26]. By intelligently adjusting the phase shifts of its reflecting elements, the IRS enhances the propagation environment between the sensors and the AP, improving both energy transfer efficiency during the downlink charging phase and uplink data transmission reliability. Our objective is to maximize the freshness and completeness of collected sensory data by jointly optimizing sensor scheduling, resource allocation, and IRS beamforming, based on channel information, energy availability, and application-specific traffic demands.

4.2.1 Time-Slotted Transmit Scheduling

Time-slotted frame structure is employed to avoid contention among sensor nodes. Each data frame is equally divided into T time slots, with the set of all time slots denoted by $\mathcal{T} \triangleq \{1, 2, \ldots, T\}$. In each time slot $t \in \mathcal{T}$, the AP's operation mode must be determined either for downlink energy transfer or uplink data transmission. Let $\psi_0(t) \in \{0, 1\}$ indicate the AP's mode selection, where $\psi_0(t) = 1$ indicates downlink energy beamforming, and $\psi_0(t) = 0$ corresponds to uplink information reception. Additionally, let $\psi_k(t) \in \{0, 1\}$ denote the uplink scheduling policy for the k-th sensor node, where $\psi_k(t) = 1$ indicates that node k is scheduled to transmit its sensed data to the AP in slot t. To prevent interference, at most one sensor node may access the channel in any given time slot, i.e.,

$$\psi_0(t) + \sum_{k \in \mathcal{K}} \psi_k(t) \leq 1, \quad \forall t \in \mathcal{T}. \tag{4.1}$$

Let $\boldsymbol{\Psi}(t) = [\psi_0(t), \psi_1(t), \ldots, \psi_K(t)]$ denote the scheduling policy at the t-th time slot, which depends on the sensor nodes' traffic demands, channel conditions, and energy status. Let $\mathcal{N} \triangleq \{1, 2, \ldots, N\}$ denote the set of IRS's reflecting elements, and let $\theta_n(t) \in (0, 2\pi]$ represent the phase shift of the n-th element in time slot t. We define the IRS's phase shift vector in slot t as $\boldsymbol{\theta}(t) = \left[e^{j\theta_1(t)}, e^{j\theta_2(t)}, \ldots, e^{j\theta_N(t)}\right] \in \mathbb{C}^{N\times 1}$. Since the IRS can configure distinct beamforming patterns for downlink and uplink transmissions, we denote the corresponding phase shift vectors as $\boldsymbol{\theta}_d(t) \triangleq \left[e^{j\theta_{d,1}(t)}, e^{j\theta_{d,2}(t)}, \ldots, e^{j\theta_{d,N}(t)}\right]$ for the downlink energy transfer phase and $\boldsymbol{\theta}_u(t) \triangleq \left[e^{j\theta_{u,1}(t)}, e^{j\theta_{u,2}(t)}, \ldots, e^{j\theta_{u,N}(t)}\right]$ for the uplink data transmission phase. The channel matrix from the AP to the IRS in time slot t is denoted by $\mathbf{G}(t) \in \mathbb{C}^{M\times N}$, where M is the number of the AP's antennas. The channel vectors from the IRS to the k-th sensor node and from the AP to the k-th sensor node are represented by $\mathbf{h}_k^r(t) \in \mathbb{C}^{N\times 1}$ and

$\mathbf{h}_k^d(t) \in \mathbb{C}^{M\times 1}$, respectively. We assume that all channels follow a quasi-static flat-fading model, i.e., they remain constant within each time slot but may vary across different slots. The AP estimates the channels via a training period at the beginning of each time slot.

4.2.2 Downlink Energy Transfer

When $\psi_0(t) = 1$, the system operates in the downlink energy transfer phase, where the IRS assists in energy transfer to sustain the sensor nodes' operation. Given the IRS's passive beamforming $\boldsymbol{\theta}_d(t)$, the effective downlink channel $\mathbf{f}_{d,k}(t)$ from the AP to the k-th sensor node is given by

$$\mathbf{f}_{d,k}(t) = \mathbf{h}_k^d(t) + \mathbf{G}(t)\boldsymbol{\Theta}_d(t)\mathbf{h}_k^r(t), \tag{4.2}$$

where $\boldsymbol{\Theta}_d(t) \triangleq \text{diag}(\boldsymbol{\theta}_d(t))$ denotes the diagonal phase shift matrix formed from the IRS's beamforming vector $\boldsymbol{\theta}_d(t)$. The phase shift matrix $\boldsymbol{\Theta}_d(t) \triangleq \text{diag}(\boldsymbol{\theta}_d(t))$ represents the IRS's passive beamforming strategy during the downlink energy transfer phase. Let $\mathbf{w}_d(t) \in \mathbb{C}^{M\times 1}$ denote the AP's active transmit beamforming for energy transfer, normalized such that $\|\mathbf{w}_d(t)\|^2 = 1$. Given the AP's transmit power p_s, the transmitted signal is $\mathbf{x}(t) = \sqrt{p_s}\mathbf{w}_d(t)s_0(t)$, where $s_0(t) \in \mathbb{C}$ is a random energy-carrying symbol with unit average power, i.e., $\mathbb{E}[|s_0(t)|^2] = 1$. The received signal at the k-th sensor node is then given by

$$y_k(t) = \mathbf{f}_{d,k}^H(t)\mathbf{x}(t) + n_k(t), \tag{4.3}$$

where $(\cdot)^H$ denotes the conjugate transpose, and $n_k(t) \sim \mathcal{CN}(0, \sigma^2)$ is additive white Gaussian noise with zero mean and unit variance σ^2. Typically the noise power during the energy transfer phase is negligible compared to the received signal strength, as the primary concern is energy harvesting rather than information decoding.

Under a linear energy harvesting model [27], the amount of energy harvested by the k-th sensor node in time slot t is expressed as

$$E_k^h(t) = \eta \cdot \mathbb{E}\left[\left|\mathbf{f}_{d,k}^H(t)\mathbf{x}(t)\right|^2\right] = \eta p_s \left|\mathbf{f}_{d,k}^H(t)\mathbf{w}_d(t)\right|^2, \tag{4.4}$$

where $\eta \in (0, 1]$ denotes the energy conversion efficiency. The contribution of the noise term $n_k(t)$ to the harvested energy is negligible and thus ignored. By optimizing the AP's transmit beamforming vector $\mathbf{w}_d(t)$ and the IRS's phase shifts $\boldsymbol{\theta}_d(t)$, the system can efficiently direct wireless energy to sensor nodes based on their channel conditions and energy demands. Specifically, the AP can steer its energy beam toward sensor nodes with insufficient energy supply. Meanwhile, the IRS's passive beamforming matrix $\boldsymbol{\Theta}_d(t)$ actively shapes the effective downlink channel $\mathbf{f}_{d,k}(t)$ to enhance the WPT for individual sensor nodes. Due to the broadcast nature of wireless signals, the energy delivered to each node depends on the joint beamforming design $(\mathbf{w}_d(t), \boldsymbol{\Theta}_d(t))$ in each time slot t. While we adopt a linear energy harvesting model in (4.4), a more practical nonlinear energy harvesting model can also be applied. In such models, the harvested power initially increases with the received signal power but eventually saturates as the input power becomes too high [28]. This behavior can be approximated using a piecewise linear model:

$$E_k^h(t) = \min\left\{\eta p_s \left|\mathbf{f}_{d,k}^H(t)\mathbf{w}_d(t)\right|^2, \ p_{\text{sat}}\right\}, \tag{4.5}$$

where p_{sat} denotes the saturation power. When the received signal power $p_s|\mathbf{f}_{d,k}^H(t)\mathbf{w}_d(t)|^2$ exceeds p_{sat}, the harvested power $E_k^h(t)$ no longer increases significantly. The proposed algorithm in the following section is developed based on the linear model in (4.4), but it can be readily extended to the piecewise linear model (4.5) with minor modifications.

4.2.3 Uplink Information Updates

We assume that the uplink channels are reciprocal to the downlink channels within each time slot, i.e., channel reciprocity holds, as commonly adopted in time division systems [18]. Let $p_k(t)$ denote the transmit power of the k-th sensor node when it is scheduled for information transmission in time slot t, i.e., $\psi_k(t) = 1$. The signal received at the AP from sensor node k is given by

$$\mathbf{y}_k(t) = \sqrt{p_k(t)}\,\mathbf{f}_{u,k}(t)s_k(t) + \mathbf{n}_k(t), \tag{4.6}$$

where $\mathbf{f}_{u,k}(t) \in \mathbb{C}^{M\times 1}$ denotes the uplink channel from the k-th sensor node to the AP, $s_k(t) \in \mathbb{C}$ is the unit-power information symbol (i.e., $\mathbb{E}[|s_k(t)|^2] = 1$), and $\mathbf{n}_k(t) \sim \mathcal{CN}(0, \mathbf{I})$ is additive white Gaussian noise with zero mean and unit covariance. Similar to the downlink channel model in (4.2), the IRS-assisted uplink channel is given by

$$\mathbf{f}_{u,k}(t) = \mathbf{h}_k^d(t) + \mathbf{G}(t)\boldsymbol{\Theta}_u(t)\mathbf{h}_k^r(t), \tag{4.7}$$

where $\boldsymbol{\Theta}_u(t) \triangleq \text{diag}(\boldsymbol{\theta}_u(t))$ represents the IRS's passive beamforming matrix during the uplink phase.

Under a time-division protocol, only one sensor node transmits in each time slot, thus avoiding mutual interference. The AP employs a receive beamforming vector $\mathbf{w}_u(t) \in \mathbb{C}^{M\times 1}$ with $\|\mathbf{w}_u(t)\|^2 = 1$ to decode the received signal. The resulting AP's SNR is

$$\gamma_k(t) = p_k(t)\left|\mathbf{f}_{u,k}^H(t)\mathbf{w}_u(t)\right|^2. \tag{4.8}$$

By aligning $\mathbf{w}_u(t)$ with the uplink channel direction, i.e., $\mathbf{w}_u(t) = \frac{\mathbf{f}_{u,k}(t)}{\|\mathbf{f}_{u,k}(t)\|}$, the AP achieves maximum receive gain and accordingly the received SNR simplifies as

$\gamma_k(t) = p_k(t)\|\mathbf{f}_{u,k}(t)\|^2$. Thus, the uplink achievable data rate for node k in slot t is then given by

$$r_k(t) = \tau_k \log_2\left(1 + p_k(t)\|\mathbf{f}_{u,k}(t)\|^2\right), \tag{4.9}$$

where $\tau_k \in (0, 1]$ denotes the fraction of the time slot allocated for uplink transmission. Given the size d_k (in bits) of sensing data, successful transmission requires that $r_k(t) \geq d_k$, i.e.,

$$\tau_k \log_2\left(1 + p_k(t)\|\mathbf{f}_{u,k}(t)\|^2\right) \geq d_k. \tag{4.10}$$

In each time slot t, when the k-th sensor node is scheduled for uplink transmission ($\psi_k(t) = 1$), its total energy consumption is given by $E_k^c(t) = \tau_k(t)\,(p_k(t) + p_c)$, where p_c denotes the constant circuit power required to maintain the node's operation (e.g., for RF front-end and baseband processing). The term $\tau_k(t)\,p_k(t)$ represents the power consumed in RF transmission, which scales linearly with both transmit power and transmission time. The transmit power $p_k(t)$ is adaptively adjusted based on the uplink channel condition $\mathbf{f}_{u,k}(t)$ to satisfy the rate requirement $r_k(t) \geq d_k$ for successful data delivery. Let $\hat{E}$ denote the maximum battery capacity of the sensor nodes, and let $E_k(t)$ represent the available energy at node k at the beginning of time slot t. Thus, the energy dynamics are given as follows:

$$E_k(t+1) = \min\left\{\left(E_k(t) - \psi_k(t)E_k^c(t)\right)^+ + \psi_0(t)E_k^h(t),\ \hat{E}\right\}, \tag{4.11}$$

where $(x)^+ \triangleq \max\{x, 0\}$ ensures non-negative residual energy, $\psi_k(t)E_k^c(t)$ is the energy consumed for uplink transmission if scheduled, and $\psi_0(t)E_k^h(t)$ is the energy harvested during downlink energy transfer (when $\psi_0(t) = 1$). To ensure successful information update, the sensor node k must have sufficient energy before transmission. Otherwise, the transmission fails and no useful data is delivered, while the allocated energy $E_k^c(t)$ is actually wasted.

4.2.4 Sensors' AoI Dynamics

Sensor nodes may increase their sampling rate when the monitored physical process changes rapidly, and reduce it when energy availability is limited. However, in this study, the physical sensing process itself is exogenous and not under our control. We focus solely on optimizing transmission scheduling and beamforming design over the wireless network. The generated sensing data at each node is assumed to be updated randomly, depending on both the dynamics of the underlying physical process and the node's current energy status. Upon acquiring new sensing data, each sensor node discards any previously cached information and stores the latest measurement. From the receiver's perspective, each successfully received update is treated as the most current information and replaces the previously stored (now obsolete) data. When sensor node k is scheduled for uplink transmission (i.e., $\psi_k(t) = 1$), it transmits the most recently cached data to the AP, which then updates its local copy upon successful reception. For each sensor node $k \in \mathcal{K}$, the caching age or data freshness depends directly on the scheduling policy $\boldsymbol{\Psi}(t)$, specifically on how frequently the node is allocated transmission opportunities. Meanwhile, the transmission time required to successfully deliver a cached update can be minimized by jointly optimizing the uplink transmit control parameters $(p_k(t), \tau_k(t))$ and the receive beamforming and IRS's phase shifts $(\mathbf{w}_u(t), \boldsymbol{\Theta}_u(t))$. This joint optimization enables efficient trade-offs between energy consumption, data freshness, and reliable communication.

Let $A_k(t)$ denote the AoI at the AP for the k-th sensor node, which characterizes the freshness of the most recently received sensing data. The AoI increases by one in each time slot if no new update is received, and resets to 1 when a successful update is delivered. Specifically, when node k is scheduled for transmission ($\psi_k(t) = 1$), the AP can update its local information only if the following conditions are met:

- The cache contains fresh data, i.e., $o_k(t) = 1$;
- The node has sufficient energy for transmission, i.e., $E_k(t) \geq E_k^c(t)$;
- The achievable uplink rate satisfies the data demand, i.e., $r_k(t) \geq d_k$.

Under these conditions, the AP successfully receives and decodes the new data at the end of slot t, and the AoI is reset in the next slot as $A_k(t+1) = 1$. Otherwise, if any of the above conditions fails, no update is received and the AoI increases as $A_k(t+1) = A_k(t) + 1$. Thus, the AoI dynamics for sensor node k are specified as follows:

$$A_k(t+1) = \begin{cases} 1, & \text{if } o_k(t) = 1,\ \psi_k(t) = 1,\ r_k(t) \geq d_k,\ E_k(t) \geq E_k^c(t), \\ A_k(t) + 1, & \text{otherwise.} \end{cases} \tag{4.12}$$

$o_k(t) \in \{0, 1\}$ indicates the status of the node's data cache. $o_k(t) = 1$ means a new data sample has been generated and cached, while $o_k(t) = 0$ indicates an empty cache. Therefore, the uplink data rate $r_k(t)$ and AoI reduction both depends on the transmit power $p_k(t)$, transmission duration $\tau_k(t)$, and the joint beamforming design $(\mathbf{w}_u(t), \boldsymbol{\Theta}_u(t))$, as defined in previous sections.

4.2.5 *Weighted AoI Minimization*

Our objective is to minimize the long-term AoI by jointly optimizing the scheduling policy $\boldsymbol{\Psi} \triangleq \{\boldsymbol{\psi}(t)\}_{t\in\mathcal{T}}$ and the beamforming strategies

$$(\mathbf{w}, \boldsymbol{\Theta}) \triangleq (\mathbf{w}_m(t), \boldsymbol{\Theta}_m(t))_{m\in\{d,u\},\, t\in\mathcal{T}}\,,$$

where d and u denote the downlink energy transfer and uplink information transmission phases, respectively. To account for heterogeneous priorities among sensor nodes, we introduce a weighted AoI metric. Let $\lambda_k > 0$ denote the priority weight for node k, reflecting the delay sensitivity of its sensing data, e.g., safety-critical monitoring in autonomous systems should have a higher λ_k. Thus, the time-averaged weighted AoI is then defined as:

$$\bar{A}(\boldsymbol{\Psi}, \mathbf{w}, \boldsymbol{\Theta}) = \limsup_{T \to \infty} \frac{1}{T} \cdot \frac{1}{K} \mathbb{E}\left[\sum_{t \in \mathcal{T}} \sum_{k \in \mathcal{K}} \lambda_k A_k(t) \right], \tag{4.13}$$

where the expectation is taken over the system dynamics (e.g., channel variations, data arrival processes), and T is the total number of time slots. We now formulate the joint scheduling and beamforming optimization problem for AoI minimization:

$$\min_{\boldsymbol{\Psi}, \mathbf{w}, \boldsymbol{\Theta}} \quad \bar{A}(\boldsymbol{\Psi}, \mathbf{w}, \boldsymbol{\Theta}), \quad \text{s.t. } (4.1) - (4.12). \tag{4.14}$$

Constraint (4.1) ensures at most one sensor is scheduled per slot, (4.11) captures the energy causality and storage dynamics, and (4.12) defines the AoI evolution. The beamforming vectors are subject to unit-norm constraints, and the IRS's phase shifts are constrained to unit modulus (e.g., ideal reflecting elements). This problem is challenging due to its long-term average objective, mixed-integer variables, non-convex constraints, and coupled dependencies across scheduling, energy, and communication resources.

Given the mode selection variable $\psi_0(t)$, the joint beamforming design $(\mathbf{w}(t), \boldsymbol{\Theta}(t))$ operates in either the downlink for energy transfer or the uplink for information transmission. In the downlink phase ($\psi_0(t) = 1$), the beamforming strategy $(\mathbf{w}_d(t), \boldsymbol{\Theta}_d(t))$ determines the amount of energy harvested by each sensor, thereby shaping their future power budgets. This energy transfer process must be jointly optimized with the uplink scheduling policy $\{\psi_k(t)\}_{k \in \mathcal{K}}$ to ensure timely information updates and improve the overall AoI performance. Given the mode selection variable $\psi_0(t)$, the joint beamforming design $(\mathbf{w}(t), \boldsymbol{\Theta}(t))$ operates in either the downlink (for wireless energy transfer) or the uplink (for information transmission). In the downlink phase ($\psi_0(t) = 1$), the beamforming strategy $(\mathbf{w}_d(t), \boldsymbol{\Theta}_d(t))$ determines the amount of energy harvested by each sensor, thereby shaping their future power budgets. This energy transfer process must be jointly optimized with the uplink scheduling policy $\{\psi_k(t)\}_{k \in \mathcal{K}}$ to ensure timely information updates and improve the overall AoI performance.

Problem (4.14) presents several key challenges. First, it is a stochastic, high-dimensional, and mixed-integer optimization problem with long-term temporal dependencies. The energy dynamics in (4.11) and the time-averaged objective in (4.13) couple decisions across time slots, making the optimal policy history-dependent. While dynamic programming could in principle solve such a problem, it becomes computationally intractable due to the curse of dimensionality arising from the large state space that includes AoI values $A_k(t)$, energy levels $E_k(t)$, channel states, and cache status $o_k(t)$ for all K nodes. Second, the data arrival process at each sensor node is generally unknown and potentially random from the AP's perspective. In an ideal scenario with perfect knowledge of future data generation instants, the AP could schedule transmissions just-in-time to minimize AoI. However, in practice, such information is unavailable. Instead, the AP must make scheduling decisions based only on causal observations, such as the current and past AoI values. For instance, a rapidly increasing AoI for a particular node may trigger a higher scheduling priority in subsequent slots, reflecting a reactive or learning-based strategy. Third, the scheduling policy $\boldsymbol{\Psi}$ is inherently combinatorial. This discrete decision space exacerbates the difficulty of joint design with continuous beamforming variables. Moreover, even with a fixed scheduling policy, the beamforming optimization must balance energy delivery in the downlink with reliable data transmission in the uplink, requiring careful adaptation to the dynamic information arrival and energy harvesting processes.

4.3 Hierarchical DRL for AoI Minimization

To overcome the aforementioned challenges, we propose a hierarchical learning framework for problem (4.14) that decomposes the joint optimization of $(\boldsymbol{\Psi}, \mathbf{w}, \boldsymbol{\Theta})$ into two coordinated modules. The overall architecture is illustrated in Fig. 4.1, comprising an outer-loop DRL agent for scheduling and an inner-loop optimization module for beamforming design. While it is possible to apply DRL to jointly learn both scheduling and beamforming policies, such an end-to-end approach would require a high-dimensional action space, including both discrete scheduling decisions and continuous beamforming variables. This typically leads to slow convergence and poor sample efficiency. Therefore, we adopt a hybrid design. The outer loop employs a model-free DRL agent to adapt the combinatorial scheduling policy $\boldsymbol{\Psi}(t)$, while the

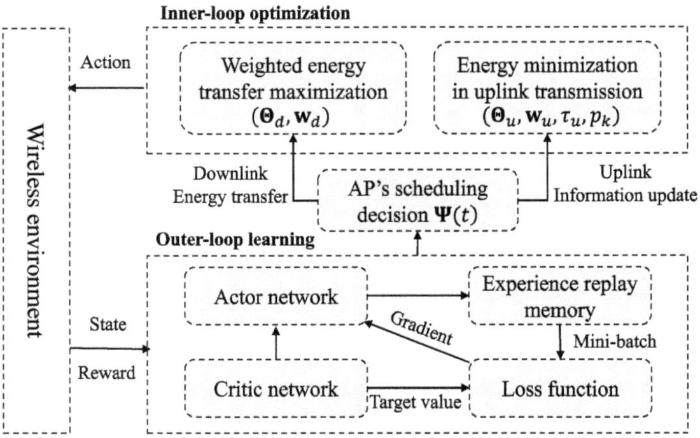

Fig. 4.1 The hierarchical DRL decouples the high-dimensional joint optimization into a two-level structure: an outer-loop DRL agent for scheduling decisions and an inner-loop optimization module for beamforming design

inner loop leverages analytical or iterative optimization methods to compute near-optimal beamforming vectors given the selected scheduling decision in the outer loop. Specifically, at each time slot t, the DRL agent first selects the scheduling action $\boldsymbol{\psi}(t)$ based on the current system state $\mathbf{s}(t)$, which includes the AoI values $\{A_k(t)\}$, energy levels $\{E_k(t)\}$, cache status $\{o_k(t)\}$, and CSI. Subsequently, the inner-loop optimization module computes the beamforming strategies $(\mathbf{w}(t), \boldsymbol{\Theta}(t))$ using AO and SDR techniques, tailored to the selected mode:

- If $\psi_0(t) = 1$, the downlink beamforming $(\mathbf{w}_d(t), \boldsymbol{\Theta}_d(t))$ is optimized to maximize the weighted sum of harvested energy across all sensor nodes.
- If $\psi_k(t) = 1$, the uplink beamforming $(\mathbf{w}_u(t), \boldsymbol{\Theta}_u(t))$ is optimized to minimize transmit power or maximize the achievable rate for node k, subject to its energy and data demand constraints.

After executing the joint action $(\boldsymbol{\psi}(t), \mathbf{w}(t), \boldsymbol{\Theta}(t))$, the AP observes the resulting system evolution, including updated AoI and energy states, and receives a cost signal (e.g., weighted AoI). This feedback is used by the DRL agent to update its policy for $\boldsymbol{\psi}(t+1)$ in the next slot. By this decomposition, the inner loop achieves computational efficiency through structured optimization, while the outer loop benefits from a significantly reduced action space, focusing only on scheduling and thus accelerating learning convergence and improving time efficiency. The proposed hybrid framework effectively balances the flexibility of learning with the efficiency of optimization, enabling scalable and adaptive control in dynamic IRS-assisted IoT networks. In the following, we first formulate and analyze the inner-loop optimization problems for joint beamforming design $(\mathbf{w}(t), \boldsymbol{\Theta}(t))$ given the scheduling decisions, and then present the outer-loop DRL approach for adaptive scheduling in Sect. 4.3.4. Given the mode selection $\psi_0(t) \in \{0, 1\}$, the AP operates in one of two modes: it either beamforms RF signals for downlink energy transfer when $\psi_0(t) = 1$, or receives sensing data from a scheduled sensor node in the uplink when $\psi_0(t) = 0$. In each case, the AP optimizes the joint active and passive beamforming strategies $(\mathbf{w}(t), \boldsymbol{\Theta}(t))$ to maximize the harvested energy in the downlink or ensure reliable data transmission in the uplink. Accordingly, we analyze the inner-loop optimization in two distinct cases, corresponding to the two operational modes of the AP.

4.3.1 Energy Minimization in Uplink Transmissions

In the t-th time slot, when the k-th sensor node is scheduled for information update (i.e., $\psi_k(t) = 1$), all other nodes must wait for future scheduling opportunities. As governed by the AoI dynamics in (4.12), the AoI of each unscheduled node increases by one, while the AoI of node k is reset to 1 if the transmission is successful ($r_k(t) \geq d_k$), or increased by one otherwise. To conserve the energy of the scheduled node for future use, we aim to minimize its energy consumption $E_k^c(t) = \tau_k(t)(p_k(t) + p_c)$ during uplink transmission, subject to the constraint of successful data delivery. This leads to the following inner-loop energy minimization problem for uplink transmission:

$$\min_{\tau_k, p_k, \mathbf{w}_u, \boldsymbol{\Theta}_u} \quad \tau_k(p_k + p_c) \tag{4.15a}$$

$$\text{s.t.} \quad \tau_k \log(1 + p_k|\mathbf{f}_k^H \mathbf{w}_u|^2) \geq d_k, \tag{4.15b}$$

$$\tau_k \in (0, 1) \text{ and } \theta_{u,n} \in (0, 2\pi], \ n \in \mathcal{N}. \tag{4.15c}$$

In the uplink transmission phase, the AP focuses on the scheduled node k, and thus only the rate constraint in (4.15b) is active. To maximize the received signal power, the AP's receive beamforming vector $\mathbf{w}_u$ is optimally aligned with the effective node k's uplink channel $\mathbf{f}_{u,k} = \mathbf{h}_k^d + \mathbf{G}_k \boldsymbol{\Theta}_u \mathbf{h}_k^r$, which combines the direct and IRS-reflected links. The optimal receive beamformer is then given by MRC $\mathbf{w}_u = \mathbf{f}_{u,k}/||\mathbf{f}_{u,k}||$. With $\mathbf{w}_u$ fixed, the passive beamforming $\boldsymbol{\Theta}_u$ aims to maximize the channel gain $\|\mathbf{f}_{u,k}\|^2$. This leads to the following optimization problem:

$$\max_{\theta_{u,n} \in (0, 2\pi],\, n=1,\ldots,N} \quad \left\| \mathbf{h}_k^d + \mathbf{G}_k \boldsymbol{\Theta}_u \mathbf{h}_k^r \right\|^2. \tag{4.16}$$

This non-convex problem can be efficiently relaxed into a SDP via rank-one relaxation as commonly applied in IRS-assisted systems [24, 29]. A rank-one solution can then be recovered via Gaussian randomization if necessary.

Given the optimized $\boldsymbol{\Theta}_u$ (and hence fixed channel gain $\|\mathbf{f}_{u,k}\|^2$), we proceed to optimize the transmission parameters (τ_k, p_k) to minimize the total energy consumption in (4.15a). Let $e_k \triangleq \tau_k p_k$ denote the energy used for RF transmission. The joint optimization of (e_k, τ_k) is formulated as follows:

$$\min_{e_k, \tau_k \in (0,1)} \quad e_k + p_c \tau_k, \tag{4.17a}$$

$$\text{s.t.} \quad \tau_k \log\left(1 + \frac{e_k}{\tau_k} ||\mathbf{f}_{u,k}^H||^2\right) \geq d_k. \tag{4.17b}$$

Problem (4.17) is jointly convex in (e_k, τ_k) and satisfies Slater's condition. Therefore, strong duality holds, and a closed-form solution (e_k^*, τ_k^*) can be derived efficiently using the Lagrangian dual method [30]. Specifically, the optimal transmit power is recovered as $p_k^* = e_k^*/\tau_k^*$. Finally, if the required energy satisfies the energy causality constraint, i.e., $e_k^* + p_c \tau_k^* \leq E_k(t)$, the transmission is successful and the AoI of node k is reset to 1. Otherwise, the transmission fails and the AoI increases accordingly.

4.3.2 AoI-Weighted Downlink Energy Maximization

When $\psi_0(t) = 1$, the AP operates in downlink WET mode, aiming to replenish the energy of all sensor nodes to support their future uplink information transmission and thereby improve the long-term time-averaged AoI performance. However, the direct impact of downlink energy transfer on future AoI is complex and non-trivial to model analytically due to the coupling across time slots, scheduling decisions, and channel dynamics. Instead, we adopt an intuitive and effective heuristic by prioritizing energy transfer to sensor nodes experiencing higher AoI, as they are in greater need of timely information updates. Specifically, if node k has a higher current AoI value $A_k(t)$, we assign it a higher priority for energy transfer. This enables the node to increase its sampling frequency and reduce transmission delays in subsequent cycles, ultimately accelerating AoI reduction.

Motivated by this intuition, we formulate the inner-loop downlink beamforming design as an AoI-weighted energy maximization problem as follows:

$$\max_{\mathbf{w}_d, \boldsymbol{\Theta}_d} \sum_{k \in \mathcal{K}} v_k |\mathbf{f}_{d,k}^H \mathbf{w}_d|^2 \tag{4.18a}$$

$$\text{s.t.} \quad E_k^c \leq E_k + E_k^h, \quad \forall k \in \mathcal{K}, \tag{4.18b}$$

$$||\mathbf{w}_d|| \leq 1 \text{ and } \theta_{d,n} \in (0, 2\pi), \quad \forall n \in \mathcal{N}, \tag{4.18c}$$

where $\mathbf{f}_{d,k}$ is the effective downlink channel from the AP to node k, $A_k(t)$ is the current AoI used as a weight, and $E_k^c = \tau_k(p_k + p_c)$ is the energy required for node k's next uplink transmission. Constraint (4.18b) ensures that the energy consumed in the next uplink transmission does not exceed the node's current energy $E_k(t)$ plus the energy E_k^h harvested in the current downlink phase. This formulation encourages the AP to focus its energy beamforming toward energy-constrained nodes that are information-stale with a higher $A_k(t)$, promoting fairness and timely status updates across the network. Specifically, for each sensor node k, we define a priority weight as follows:

$$v_k = A_k(t) + \alpha_k E_k^{-1}(t),$$

which increases with the current AoI $A_k(t)$ and decreases with the available energy $E_k(t)$. By incorporating v_k as the weighting coefficient in the objective (4.18a), the AP directs more energy to nodes that are most in need of timely updates and are at risk of transmission failure due to energy shortage. The parameter $\alpha_k > 0$ controls the trade-off between AoI and energy sensitivity for node k. A larger α_k indicates that the node is more vulnerable to energy insufficiency, and thus should be prioritized in downlink energy transfer even if its AoI is not the highest.

Furthermore, we adopt a conservative and proactive energy management policy by ensuring that every sensor node has sufficient energy to transmit successfully if it is scheduled in the next slot. To this end, we impose constraint (4.18b), $E_k^c \leq E_k(t) + E_k^h$, which guarantees that the energy required for uplink transmission $E_k^c = \tau_k(p_k + p_c)$ does not exceed the sum of the node's current energy $E_k(t)$ and the energy E_k^h harvested in the current downlink phase. This constraint acts as a one-step look-ahead safety mechanism. Without it, the beamforming optimization might neglect certain nodes, leaving them with insufficient energy. If such a node is subsequently scheduled by the outer-loop DRL agent, its transmission would fail, leading to an unnecessary

AoI increase and degraded performance. By enforcing (4.18b), we eliminate this risk and enhance system robustness, ensuring that the DRL agent can schedule any node in the next slot without being constrained by energy availability. Thus, the combination of the AoI-weighted objective and the energy sufficiency constraints enables a balanced and reliable energy provisioning to support the long-term AoI performance improvement.

4.3.3 Alternating Optimization and Approximation

Given the uplink beamforming $(\mathbf{w}_u, \boldsymbol{\Theta}_u)$ and transmission parameters $(\tau_k, p_k)_{k\in\mathcal{K}}$, the downlink beamforming $(\mathbf{w}_d, \boldsymbol{\Theta}_d)$ in problem (4.18) is solved using the AO framework, as commonly adopted in IRS-assisted systems [29, 31]. In particular, the joint optimization of $(\mathbf{w}_d, \boldsymbol{\Theta}_d)$ is decomposed into two subproblems: (i) passive beamforming $\boldsymbol{\Theta}_d$ with fixed $\mathbf{w}_d$, and (ii) active beamforming $\mathbf{w}_d$ with fixed $\boldsymbol{\Theta}_d$.

4.3.3.1 AO for Joint Downlink Beamforming

We first optimize the IRS's phase shifts $\boldsymbol{\Theta}_d$, which affect the objective (4.18a) and constraints through the effective downlink channel $\mathbf{f}_{d,k}$. Define $\mathbf{F}_k \triangleq \mathbf{G}_k \text{diag}(\mathbf{h}_k^r)$, so that we have $\mathbf{f}_{d,k} = \mathbf{h}_k^d + \mathbf{F}_k \boldsymbol{\theta}_d$. To handle the quadratic form $|\mathbf{f}_{d,k}^H \mathbf{w}_d|^2$, we introduce an auxiliary variable $\bar{\boldsymbol{\theta}}_d = [\boldsymbol{\theta}_d^T, \zeta]^T$ with $\zeta \in \mathbb{C}$, $|\zeta| = 1$, and define the extended matrix

$$\mathbf{R}_k = \begin{bmatrix} \mathbf{F}_k^H \mathbf{w}_d \mathbf{w}_d^H \mathbf{F}_k & \mathbf{F}_k^H \mathbf{w}_d \mathbf{w}_d^H \mathbf{h}_k^d \\ (\mathbf{h}_k^d)^H \mathbf{w}_d \mathbf{w}_d^H \mathbf{F}_k & 0 \end{bmatrix}.$$

Then, $|\mathbf{f}_{d,k}^H \mathbf{w}_d|^2 = \bar{\boldsymbol{\theta}}_d^H \mathbf{R}_k \bar{\boldsymbol{\theta}}_d + \|\mathbf{h}_k^d\|^2$. Since the constant term does not affect optimization, we focus on $\bar{\boldsymbol{\theta}}_d^H \mathbf{R}_k \bar{\boldsymbol{\theta}}_d$. Let $\boldsymbol{\Phi}_d = \bar{\boldsymbol{\theta}}_d \bar{\boldsymbol{\theta}}_d^H \succeq 0$ and we relax the rank-one constraint. The harvested energy at node k is $E_k^h = \eta\, |\mathbf{f}_{d,k}^H \mathbf{w}_d|^2$. Thus, the energy sufficiency constraint (4.18b), $E_k^c \le E_k + E_k^h$, becomes

$$|\mathbf{f}_{d,k}^H \mathbf{w}_d|^2 \ge \eta^{-1}(E_k^c - E_k).$$

The rank-one relaxation yields $\text{Tr}\,(\mathbf{R}_k \boldsymbol{\Phi}_d) \ge \eta^{-1}(E_k^c - E_k)$, where $\text{Tr}(\cdot)$ denotes the matrix trace. Additionally, the unit-modulus constraint on IRS implies $\boldsymbol{\Phi}_d(n, n) = 1$ for $n = 1, \ldots, N + 1$. The resulting SDP becomes:

$$\max_{\boldsymbol{\Phi}_d \succeq 0} \sum_{k\in\mathcal{K}} v_k \, \text{Tr}\,(\mathbf{R}_k \boldsymbol{\Phi}_d) \tag{4.19a}$$

$$\text{s.t.}\ \text{Tr}\,(\mathbf{R}_k \boldsymbol{\Phi}_d) \ge \eta^{-1}(E_k^c - E_k)^+, \quad \forall k \in \mathcal{K}, \tag{4.19b}$$

$$\boldsymbol{\Phi}_d(n, n) = 1, \quad n = 1, \ldots, N + 1, \tag{4.19c}$$

Algorithm 1 AO for Joint Downlink Beamforming

Input: The channel information $\{\mathbf{h}_k^d(t), \mathbf{G}_k(t), \mathbf{h}_k^r(t)\}$, AoI state A_k, energy state E_k, and energy demand E_k^c of each sensor node $k \in \mathcal{K}$

Initialize: a feasible beamforming strategy $(\mathbf{w}_d, \boldsymbol{\Theta}_d)$, $\tau \leftarrow 0$

$v_k \leftarrow A_k + \alpha_k E_k^{-1}$

$E_d^{(\tau)} \leftarrow 0$, $E_d^{(\tau+1)} \leftarrow \sum_{k\in\mathcal{K}} v_k |\mathbf{f}_{d,k}^H \mathbf{w}_d|^2$

while $||E_d^{(\tau+1)} - E_d^{(\tau)}|| \geq \epsilon$ **do**

  $\tau \leftarrow \tau + 1$

  Solve SDP (4.19) by the interior-point algorithm

  Extract the rank-one passive beamformer $\boldsymbol{\Theta}_d$

  Update $\mathbf{w}_d$ with the fixed $\boldsymbol{\Theta}_d$

  $E_d^{(\tau)} \leftarrow E_d^{(\tau+1)}$

  $E_d^{(\tau+1)} \leftarrow \sum_{k\in\mathcal{K}} v_k |\mathbf{f}_{d,k}^H \mathbf{w}_d|^2$

end while

where $(x)^+ = \max(x, 0)$ ensures feasibility when $E_k \geq E_k^c$. Problem (4.19) is a convex SDP and can be efficiently solved using interior-point methods [32].

With $\boldsymbol{\Theta}_d$ and thus $\mathbf{f}_{d,k}$ fixed, we optimize the active beamformer $\mathbf{w}_d$. We similarly apply SDR by defining $\mathbf{W}_d = \mathbf{w}_d \mathbf{w}_d^H \succeq 0$, relaxing the rank-one constraint. Then, $|\mathbf{f}_{d,k}^H \mathbf{w}_d|^2 = \mathrm{Tr}(\mathbf{f}_{d,k}\mathbf{f}_{d,k}^H \mathbf{W}_d)$, and the power constraint $\|\mathbf{w}_d\|^2 \leq 1$ becomes $\mathrm{Tr}(\mathbf{W}_d) \leq 1$. The energy constraint (4.18b) is similarly reformulated. The resulting SDP in $\mathbf{W}_d$ is analogous to (4.19) and can be solved efficiently. After solving each SDP, we extract a rank-one solution via Gaussian randomization method. The AO procedure is shown in Algorithm 1, which alternates between optimizing $\boldsymbol{\Phi}_d$ (or $\boldsymbol{\theta}_d$) and $\mathbf{W}_d$ (or $\mathbf{w}_d$) until the objective value converges to a stable point.

4.3.3.2 Approximation with Reduced Complexity

The inner-loop optimization must be computationally efficient, minimizing both processing delay and resource overhead. This efficiency is critical for enabling real-time execution, especially when the inner-loop optimization is integrated within the outer-loop DRL framework that requires rapid and repeated evaluation at each time step. While the AO method in Algorithm 1 guarantees convergence, it may incur high latency due to repeated solutions of the SDP (4.19) across multiple iterations. The number of required iterations depends on initialization and channel conditions, making worst-case delay unpredictable. To address this, we propose a low-complexity approximation strategy that avoids iterative AO by fixing the active beamforming $\mathbf{w}_d$ based on a heuristic channel alignment principle, and optimizing only the passive beamforming strategy $\boldsymbol{\Theta}_d$. This reduces the overall complexity of solving multiple SDPs per slot, significantly improving computational efficiency. Our approximation proceeds in three steps:

4.3.3.3 Step 1: Idealized Energy Maximization

We first consider an optimistic relaxation in which the AP can simultaneously align $\mathbf{w}_d$ to all effective downlink channels $\mathbf{f}_{d,k}$, a scenario only possible if all channels are co-directional. Under this idealization, the received energy at node k becomes proportional to $\|\mathbf{f}_{d,k}\|^2$, leading to the relaxed problem:

$$\max_{\boldsymbol{\Theta}_d} \sum_{k\in\mathcal{K}} v_k \|\mathbf{f}_{d,k}\|^2, \quad \text{s.t.} \quad (4.18b) \text{ and } (4.18c), \tag{4.20}$$

where $\mathbf{f}_{d,k} = \mathbf{h}_k^d + \mathbf{G}_k\boldsymbol{\Theta}_d\mathbf{h}_k^r$. This problem depends only on $\boldsymbol{\Theta}_d$ and can be efficiently solved via SDR using the same method as for (4.16). However, this relaxation overestimates the achievable energy transfer, since perfect multi-user beamforming alignment is generally infeasible in practice.

4.3.3.4 Step 2: Select Robust Beamforming Vector

To obtain a feasible $\mathbf{w}_d$, we evaluate the channel gains $\|\mathbf{f}_{d,k}\|^2$ under the solution $\boldsymbol{\Theta}_d$ to the relaxed problem (4.20) and sort sensor nodes by their channel gains. We then select the weakest effective channel:

$$\mathbf{f}_{d,k}^m = \arg\min_{k\in\mathcal{K}} \|\mathbf{f}_{d,k}\|.$$

Thus, the AP's active beamforming vector is aligned to the weakest channel as $\mathbf{w}_d = \|\mathbf{f}_{d,k}^m\|^{-1}\mathbf{f}_{d,k}^m$, which ensures maximum energy transfer to the most disadvantaged node, promoting fairness and robustness in energy transfer.

4.3.3.5 Step 3: Feasible IRS Optimization

With $\mathbf{w}_d$ fixed, we re-optimize $\boldsymbol{\Theta}_d$ by solving the original SDP (4.19). This yields a feasible solution $(\mathbf{w}_d, \boldsymbol{\Theta}_d)$ that satisfies all constraints and provides a lower bound on the optimal objective of (4.18).

In summary, this approximation replaces the iterative AO algorithm by a two-SDP procedure, effectively decoupling the optimization of $\mathbf{w}_d$ and $\boldsymbol{\Theta}_d$. This eliminates the need for repeated iterations, significantly improving the computational efficiency of the inner loop optimization while ensuring predictable time overhead. Step 1 maximizes a relaxed objective and yields an initial IRS configuration $\boldsymbol{\Theta}_d$ that enhances the effective downlink channels for all nodes. In step 2, we align the active beamforming vector $\mathbf{w}_d$ to the weakest effective channel (identified in step 2). This conservative design prioritizes the most disadvantaged sensor node, ensuring robust energy delivery even under poor channel conditions. In step 3, with $\mathbf{w}_d$ fixed, we solve a final SDP to re-optimize $\boldsymbol{\Theta}_d$, resulting in a feasible solution that provides a lower bound on the

Algorithm 2 Two-SDP Approximation for $(\mathbf{w}_d, \boldsymbol{\Theta}_d)$

Input: The channel information $\{\mathbf{h}_k^d(t), \mathbf{G}_k(t), \mathbf{h}_k^r(t)\}$, AoI state A_k, energy state E_k, and energy demand E_k^c of each sensor node $k \in \mathcal{K}$

1: $v_k \leftarrow A_k + \alpha_k E_k^{-1}$
2: Solve relaxed problem (4.20) in the optimistic case
3: $\mathbf{f}_{d,k}^m \leftarrow \arg\min_{k\in\mathcal{K}} ||\mathbf{f}_{d,k}||$
4: $\mathbf{w}_d \leftarrow \|\mathbf{f}_{d,k}^m\|^{-1}\mathbf{f}_{d,k}^m$
5: Solve problem (4.19) with the fixed $\mathbf{w}_d$
6: Extract the passive beamforming strategy $\boldsymbol{\Theta}_d$

original objective. By solving only two SDPs per time slot, the proposed approximation algorithm achieves a favorable trade-off between performance and complexity. This computational efficiency enables timely execution of the inner-loop optimization, making the overall hierarchical learning framework both scalable and practical for real-time networks.

4.3.4 Outer-Loop Model-Free Learning for Scheduling

The outer-loop scheduling policy is optimized using model-free DRL, in which the AP learns to make intelligent decisions through continuous interaction with the dynamic network environment. We formulate the scheduling problem into MDP, defined by the tuple $(\mathcal{S}, \mathcal{A}, \mathcal{R})$, where $\mathcal{S}$ is the state space, $\mathcal{A}$ is the action space, and $\mathcal{R}$ is the reward function. At each decision epoch t, the system state observed by the AP is denoted as $\mathbf{s}_t \in \mathcal{S}$. It captures the time-varying information freshness and energy status of all sensor nodes:

$$\mathbf{s}_t = \big(\mathbf{A}(t),\ \mathbf{E}(t)\big),$$

where $\mathbf{A}(t) \triangleq [A_1(t), A_2(t), \ldots, A_K(t)]$ is the vector of AoI values, and $\mathbf{E}(t) \triangleq [E_1(t), E_2(t), \ldots, E_K(t)]$ represents the current energy levels. By definition, $A_k(t) \geq 1$ evolves over time, increasing by one each slot if node k is not scheduled, and resetting upon successful update reception. The energy level $E_k(t)$ is bounded within $[0, \hat{E}]$, reflecting the finite battery capacity of each sensor node. The action at time t is the scheduling decision $\mathbf{a}_t \in \mathcal{A}$, given by:

$$\mathbf{a}_t = \{\psi_0(t), \psi_1(t), \ldots, \psi_K(t)\} \in \{0, 1\}^{K+1},$$

where $\psi_k(t) = 1$ indicates that node k is scheduled in slot t, and $\psi_0(t) = 1$ corresponds to an idle state. The action must satisfy the scheduling constraint (4.1), ensuring that at most one node is served per slot. Given $\mathbf{a}_t$, the inner-loop optimization computes the beamforming strategies $(\mathbf{w}, \boldsymbol{\Theta})$ and transmit parameters (τ_k, p_k), enabling energy transfer and data collection. This in turn updates the AoI and energy states for the next time step. The reward quantifies the immediate cost of taking action $\mathbf{a}_t$ in state $\mathbf{s}_t$, guiding the agent toward long-term performance improvement. To align with the AoI minimization in (4.14), the reward at time t is defined as follows:

$$v_t(\mathbf{s}_t, \mathbf{a}_t) = -\frac{1}{K|\mathcal{H}_t|} \sum_{\tau \in \mathcal{H}_t} \sum_{k \in \mathcal{K}} \lambda_k A_k(\tau), \tag{4.21}$$

where $\mathcal{H}_t \triangleq \{t - t_o + 1, \ldots, t\}$ is a sliding window of the most recent t_o time slots, and $\lambda_k > 0$ is a priority weight reflecting the importance of node k's information. This reward function represents the negative weighted average AoI over the observation window $\mathcal{H}_t$, encouraging the DRL agent to maintain low and stable information age across the network.

DRL methods such as DQN and policy gradient algorithms have been proven effective in solving complex MDPs in large-scale wireless networks by leveraging DNNs to approximate either the value function or the policy. In this case study, we adopt the PPO algorithm [33], a state-of-the-art on-policy actor-critic method known for its excellent sample efficiency, training stability, and ease of implementation. PPO achieves a favorable trade-off between performance and complexity, making it well-suited for real-time scheduling in energy-constrained IoT systems with dynamic channel and energy conditions. To adapt the outer-loop scheduling policy within the proposed hierarchical learning framework, we integrate PPO as detailed in Algorithm 3. At the start of training, the DNN parameters $\boldsymbol{\omega}$ of the policy network are randomly initialized. In each time step t of a learning episode, the AP observes the current system state $\mathbf{s}_t = (\mathbf{A}(t), \mathbf{E}(t))$ and uses the policy network to select a scheduling action $\boldsymbol{\Psi}(t) = \{\psi_k(t)\}_{k=0}^{K}$, as outlined in line 4 of Algorithm 3. Given the scheduling decision $\boldsymbol{\Psi}(t)$, the AP performs inner-loop optimization to compute the joint beamforming strategies $(\mathbf{w}(t), \boldsymbol{\Theta}(t))$. Specifically, if a sensor node is scheduled for uplink data transmission, the AP optimizes the receive beamformer $\mathbf{w}(t)$ and IRS's phase shifts $\boldsymbol{\Theta}(t)$ to maximize the uplink sum rate. Conversely, during downlink energy transfer (when $\psi_0(t) = 1$), the AP designs the energy beamformer $\mathbf{w}_d(t)$ and reflection matrix $\boldsymbol{\Theta}_d(t)$, computed either via the iterative AO-based Algorithm 1 or the low-complexity approximation in Algorithm 2, as shown in lines 5–6 of Algorithm 3. Once both outer-loop and inner-loop decisions are determined, the joint action $(\boldsymbol{\Psi}(t), \mathbf{w}(t), \boldsymbol{\Theta}(t))$ is executed in the environment. The system transitions to the next state, and the corresponding reward $v_t(\mathbf{s}_t, \mathbf{a}_t)$, defined in (4.21), is computed and stored in the experience replay buffer (lines 7–8). The policy network is updated periodically using mini-batch gradient descent on sampled experiences from the replay buffer (lines 11–13). For benchmarking, we also implement a DQN-based scheduler. Numerical results presented in Sect. 4.4 demonstrate that the PPO-based Algorithm 3 achieves faster convergence and consistently lower average AoI compared to DQN, validating its superior learning efficiency and robustness in this control task.

Algorithm 3 PPO-based Hierarchical Scheduling and Transmit Control

Initialize: Target policy π_{ω} and behavior policy π_{ω}^{o},
Initialize: $t \leftarrow 0$, $E_k(0) \leftarrow B$, $A_k(t) \leftarrow 0$
for Episode $= \{1, 2, \ldots, \text{MAX} = 3000\}$ **do**
  while $t \neq T$ **do**
    Observe the system state $(\mathbf{A}(t), \mathbf{E}(t))$
    Choose the outer-loop action $\boldsymbol{\Psi}(t)$ for scheduling
    case $\psi_0(t) = 0$: optimize uplink data transmission in (4.15)
    case $\psi_0(t) = 1$: optimize downlink energy transfer in (4.18)
    Execute joint action $\mathbf{a}(t) \triangleq (\boldsymbol{\Psi}(t), \mathbf{w}(t), \boldsymbol{\Phi}(t))$, evaluate $v_t(\mathbf{s}_t, \mathbf{a}_t)$
    Observe the next state $\mathbf{s}_{t+1}$ and buffer the transition $(\mathbf{s}_t, \mathbf{a}_t, v_t, \mathbf{s}_{t+1})$
    $t \leftarrow t + 1$
  end while
  Take mini-batch form the experience replay buffer
  DNN training to update policy parameter $\boldsymbol{\omega}$
  Update behavior policy $\pi_{\omega}^{o} \leftarrow (1 - \mu)\pi_{\omega}^{o} + \mu\pi_{\omega}$
end for

4.4 Numerical Evaluation

In this section, we present simulation results to evaluate the performance of the proposed hierarchical learning framework for wireless-powered and IRS-assisted sensor networks. We consider a three-dimensional deployment where the AP and the IRS are located at $(200, -200, 0)$ and $(0, 0, 0)$ in meters, respectively. The K sensor nodes are uniformly distributed within a rectangular area $[5, 35] \times [-35, 35]$ on the (x, y)-plane at height $z = -20$ m. We adopt a realistic channel model to account for both large-scale path loss and small-scale fading. The direct channel from the AP to sensor node k, denoted $\mathbf{h}_k^d(t)$, follows a Rayleigh fading model $\mathbf{h}_k^d(t) = \beta_{0,k}\tilde{\mathbf{h}}_k^d(t)$, where $\tilde{\mathbf{h}}_k^d(t) \sim \mathcal{CN}(0, \mathbf{I})$ represents the small-scale fading component, and $\beta_{0,k}$ is the square-root of the path loss given by $\beta_{0,k} = 10^{-\frac{32.6+36.7\log_{10}(d_k^{\text{AS}})}{20}}$, with d_k^{AS} being the distance (in meters) between the AP and sensor node k. The exponent 36.7 corresponds to an urban outdoor environment. A similar channel model is employed in [34]. Similarly, the IRS-sensor channel $\mathbf{h}_k^r(t)$ and the AP-IRS channel $\mathbf{G}(t)$ are modeled with Rician fading to account for the presence of a dominant LoS component. All channels are assumed to vary slowly over time, remaining constant within each time slot but potentially changing across slots to reflect user mobility or environmental dynamics. The other system parameters are summarized as follows: The AP is equipped with 4 antennas and transmits at a power of 30 dBm. The energy conversion efficiency for WPT is set to 0.9, and the noise power is -75 dBm. Each sensor node has 5 Kbits of data to transmit. The DNN used in the learning framework has 3 hidden layers with 64 neurons per layer, employing Tanh and Softmax activation functions, and is trained using the Adam optimizer. The actor and critic networks are updated with learning rates of 0.0005 and 0.001, respectively, and a reward discount factor of 0.99 is applied to prioritize long-term performance.

4.4.1 Learning Efficiency and Convergence

Figure 4.2 compares the learning performance of the proposed hierarchical PPO algorithm (Algorithm 3) against three benchmark schemes: (i) a hierarchical DQN approach, (ii) a conventional end-to-end PPO (denoted as All-in-One PPO), and (iii) a Two-Step PPO variant where both outer-loop scheduling and inner-loop beamforming are learned via PPO without analytical guidance. The All-in-One PPO treats the joint decision of scheduling $\boldsymbol{\Psi}(t)$ and beamforming $(\mathbf{w}(t), \boldsymbol{\Theta}(t))$ as a single action in a mixed discrete-continuous action space. As shown by the black solid curve, this approach suffers from poor convergence and unstable training. The high dimensionality and heterogeneity of the action space lead to inefficient exploration and delayed policy improvement, resulting in consistently low reward values throughout training. The hierarchical DQN (black dotted curve) decouples scheduling from beamforming but relies on a value-based DQN with discrete action selection. While this reduces complexity, it exhibits significant training instability and fails to converge to a high-performing policy. In contrast, the proposed hierarchical PPO (red solid curve) achieves both faster convergence and superior asymptotic performance. By decomposing the problem, the outer-loop PPO agent operates in a reduced discrete action space for scheduling only, while the inner-loop optimization via Algorithm 1 provides analytically grounded and high-quality beamforming solutions. This synergy allows the outer-loop policy to receive more informative rewards, accelerating learning and guiding the agent toward optimal scheduling decisions. We further evaluate the Two-Step PPO, where both scheduling and beamforming are optimized using

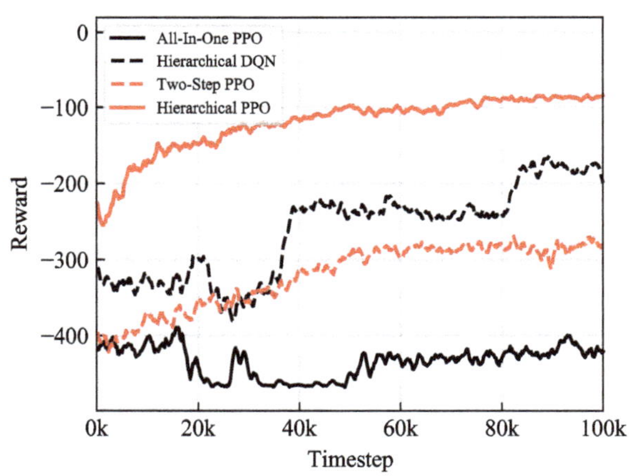

Fig. 4.2 Hierarchical PPO outperforms three benchmarks. All-in-One PPO suffers from instability due to high-dimensional mixed action space. Hierarchical DQN reduces complexity but lacks convergence. Two-step PPO decomposes actions but learns both parts via PPO, leading to suboptimal performance. In contrast, hierarchical PPO combines outer-loop PPO for scheduling with inner-loop analytical beamforming optimization, enabling faster convergence, higher rewards, and improved performance

separate PPO agents without inner-loop analytical optimization. While this approach converges better than All-in-One PPO, benefiting from action space decomposition, it still underperforms the proposed framework. This highlights the critical role of optimization-driven learning, replacing heuristic or learned beamforming with model-based inner-loop optimization provides superior training signals, leading to more effective outer-loop policy learning. In summary, Fig. 4.2 demonstrates that the proposed hierarchical PPO framework not only mitigates the curse of dimensionality but also leverages analytical domain knowledge to achieve faster convergence and higher reward, leading to a lower AoI and improved system performance.

We further compare the computational efficiency and performance of the AO-based iterative Algorithm 1 and the simplified Algorithm 2 for inner-loop beamforming optimization. As shown in Fig. 4.3a, the execution time of both algorithms increases with the number of IRS elements N and the number of sensor nodes K, which is expected due to the growing dimensionality of the beamforming and phase-shift variables. However, Algorithm 2 achieves a significant reduction in runtime by avoiding iterative updates, particularly in large-scale setups with massive IRS arrays. For instance, when $N = 80$ and $K = 6$, Algorithm 2 reduces the average computation time by almost 70% compared to Algorithm 1. Despite the reduced complexity, the long-term reward remains nearly identical between the two methods, as illustrated in Fig. 4.3b, indicating that the closed-form or heuristic approximations in Algorithm 2 effectively capture the essential beamforming and reflection gains. This close-to-optimal performance with substantially lower latency makes Algorithm 2 highly favorable for real-time implementation in resource-constrained IoT systems. Therefore, we recommend the simplified algorithm in practical deployments where timely decision-making is critical, without sacrificing significant performance.

4.4.2 Comparison with Baseline Scheduling Policies

To benchmark the performance of the proposed hierarchical PPO algorithm (Algorithm 3), we consider two conventional scheduling policies: (i) Round-Robin (ROR) and (ii) Max-AoI-First (MAF). The ROR policy cyclically schedules sensor nodes in a fixed order, ensuring fairness but ignoring real-time channel and energy conditions. Within each scheduling slot, the AP jointly optimizes active beamforming $\mathbf{w}(t)$ and IRS's phase shifts $\boldsymbol{\Theta}(t)$ to maximize the achievable uplink rate for the scheduled node. In contrast, the MAF policy is opportunistic. At each time slot, the AP selects the sensor node with the highest current AoI, prioritizing those with the most outdated status [35]. This heuristic aims to minimize peak AoI but does not account for energy availability or channel quality.

Both baseline policies employ the same energy harvesting protocol. The AP initiates downlink energy transfer only when the scheduled node's battery level falls below a predefined threshold, ensuring minimal interference with uplink transmissions. Figure 4.4a illustrates the average AoI performance as the number of sensor nodes K increases, with all algorithms operating under identical network topologies.

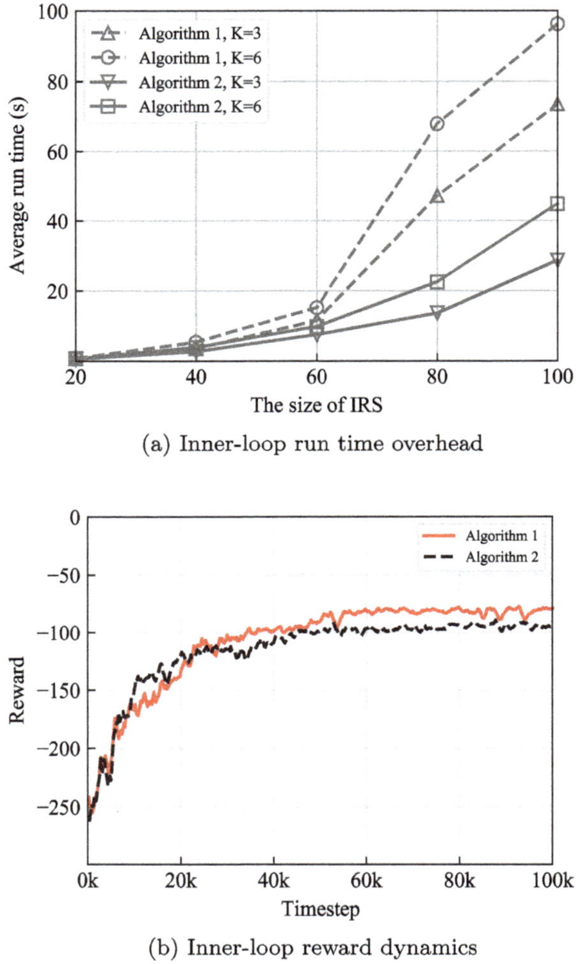

Fig. 4.3 Algorithm 2 converges to near-identical long-term reward with significantly lower computation time compared to Algorithm 1, making it highly efficient for practical deployment

When K is small, all schemes achieve relatively low AoI due to reduced contention and sufficient energy availability. In this regime, MAF outperforms ROR by proactively serving stale nodes, demonstrating the benefit of AoI-aware scheduling. However, as K grows, the limitations of the baselines become evident. ROR fails to adapt to dynamic AoI and channel conditions, while MAF may select nodes with poor channel gain or insufficient energy, leading to transmission failures or low data quality. In contrast, Algorithm 3 dynamically balances AoI, channel state, and energy levels through its learned policy, enabling intelligent joint scheduling and resource allocation. Consequently, the proposed algorithm consistently achieves lower average AoI across all network scales, with increasing performance gains as K increases. This

Fig. 4.4 Algorithm 3 achieves superior performance over ROR and MAF. While ROR lacks adaptability and MAF ignores channel and energy states, Algorithm 3 learns a policy to proactively balance AoI, channel quality, and energy availability, achieving lowest AoI and ensuring fairness among nodes

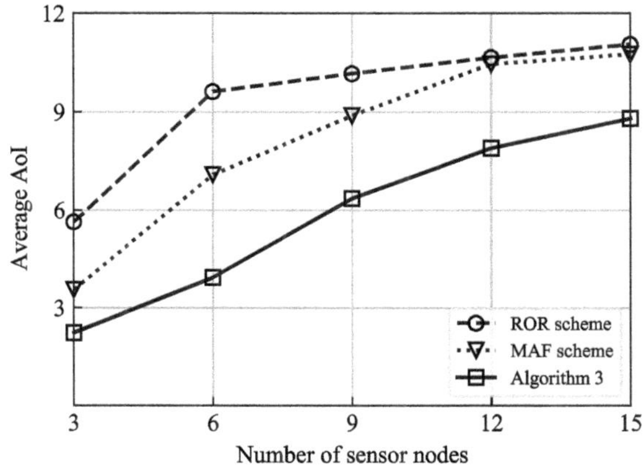

(a) Minimum AoI with our algorithm

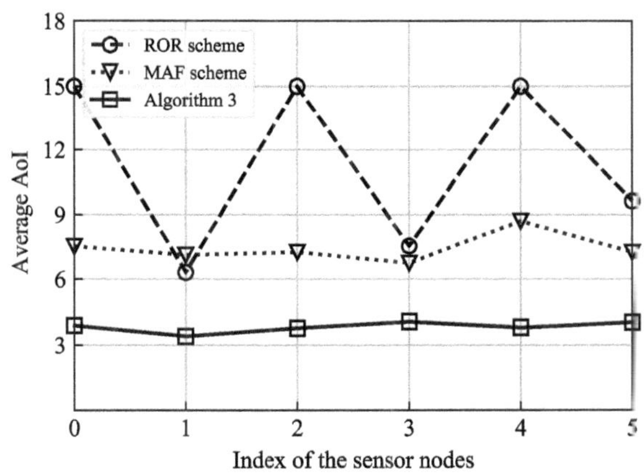

(b) Fairness among different nodes

highlights the superiority of model-free DRL in capturing complex system dynamics and optimizing long-term information freshness in large-scale IRS-assisted sensor networks.

Figure 4.4b evaluates the fairness of the scheduling policies by comparing the average AoI across individual sensor nodes. For a fair comparison, all nodes are assigned identical weight parameters $\lambda_k = \lambda$ in the weighted AoI objective (4.13). Algorithm 3 not only achieves the lowest average AoI for all nodes but also ensures highly consistent performance across the network. The small variance in nodes' AoI indicates that the learned policy effectively balances information freshness among all sensors, avoiding persistent neglect of any particular node. In contrast, the ROR scheme exhibits significantly higher deviation in AoI across nodes. Although ROR provides temporal fairness in access opportunities, it fails to adapt to dynamic data generation patterns, channel fluctuations, and energy availability. Consequently, some nodes experience unnecessarily long update delays despite being scheduled, while others may be scheduled when they have no fresh data to transmit, leading to inefficient resource usage and uneven AoI distribution. Interestingly, the MAF policy also achieves relatively low AoI variance, outperforming ROR in fairness despite being greedy. This is because MAF inherently prevents any single node from experiencing excessively stale status updates. Whenever a node's AoI becomes dominant, it is prioritized for transmission. Thus, MAF acts as an implicit load-balancing mechanism that caps the maximum AoI across the network. Nevertheless, Algorithm 3 surpasses both baselines by combining the benefits of adaptability and fairness. It learns to anticipate traffic patterns, exploit favorable channel conditions, and proactively schedule nodes before their AoI becomes critical, resulting in uniformly low and stable information age for all users.

4.4.3 AoI Reduction Using a Larger IRS

The deployment of an IRS enhances both downlink WPT and uplink information transmission, thereby implicitly improving the system's AoI performance. To evaluate the impact of the IRS, Fig. 4.5 illustrates the transmission failure rate and average AoI as the IRS's size increases from 20 to 100 elements. The transmission failure rate, computed over 30 K time slots and shown in Fig. 4.5a, decreases significantly with larger IRS sizes. This is due to the IRS's ability to reconfigure wireless channels, improving both channel strength and reliability for uplink transmission.

As a result, the success probability of status updates increases, leading to lower AoI in the long run. However, AoI improvement eventually saturates, as observed in Fig. 4.5b. The performance gap between the proposed method and baseline schemes diminishes with increasing IRS size, indicating a shift in the performance bottleneck. With a sufficiently large IRS, channel conditions become favorable and transmission delays are minimized. Consequently, scheduling delay rather than link reliability becomes the dominant factor limiting AoI performance. This underscores the need for intelligent and adaptive scheduling to fully exploit the benefits of large-scale IRS deployments.

4.4.4 AoI and Energy Trade-Off

In this section, we investigate the trade-off between AoI and energy consumption in an energy-constrained sensor network. To enable this trade-off in the DRL framework, we revise the DRL agent's reward function as:

Fig. 4.5 **a** Larger IRS sizes reduce transmission failures and improve AoI by enhancing channel conditions. **b** Performance gains become saturated as improved link quality alone is insufficient for further AoI reduction in IRS-assisted sensor networks

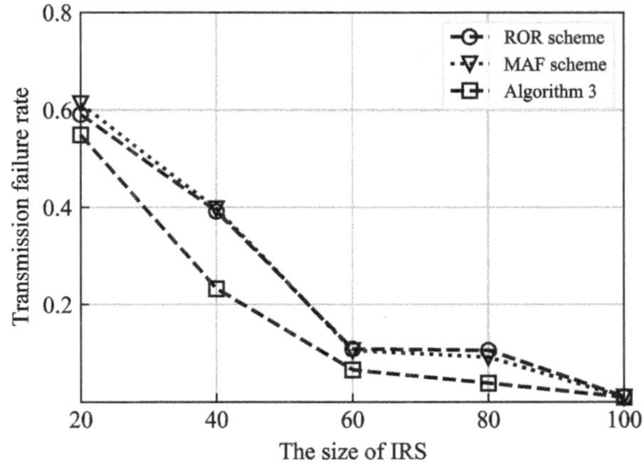

(a) IRS' size vs. failure rate

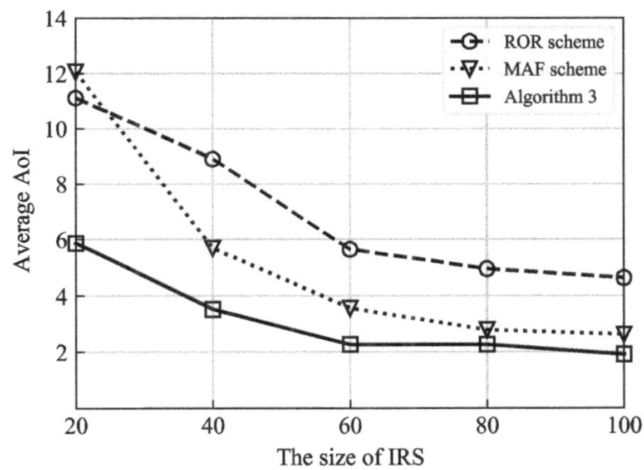

(b) Larger-size IRS reduces AoI

$$v_t(\mathbf{s}_t, \mathbf{a}_t) = -\frac{1}{K|\mathcal{H}_t|} \sum_{\tau \in \mathcal{H}_t} \sum_{k \in \mathcal{K}} \Big(\lambda_k A_k(\tau) + \lambda_0 E_k^c(\tau) \Big),$$

where λ_0 is a tunable parameter that balances AoI minimization and energy conservation. A smaller λ_0 prioritizes timely updates, encouraging aggressive transmission behavior that may lead to energy depletion. In contrast, a larger λ_0 makes the agent more energy-aware, tolerating higher AoI to preserve battery levels and avoid energy outages. As shown in Fig. 4.6a, varying λ_0 leads to distinct performance trends. When $\lambda_0 = 0$, energy cost is ignored, and the average AoI is reduced by 53.8% compared to the case with $\lambda_0 = 2000$. This demonstrates the significant impact of energy weighting on AoI performance and highlights the importance of adaptive control in energy-constrained sensor networks. We further evaluate the algorithm's

Fig. 4.6 Energy-aware AoI minimization by the proposed heterogeneous PPO framework, effectively balancing AoI and energy consumption through a tunable reward function. We set $N = 80$ and $K = 3$ in the experiment

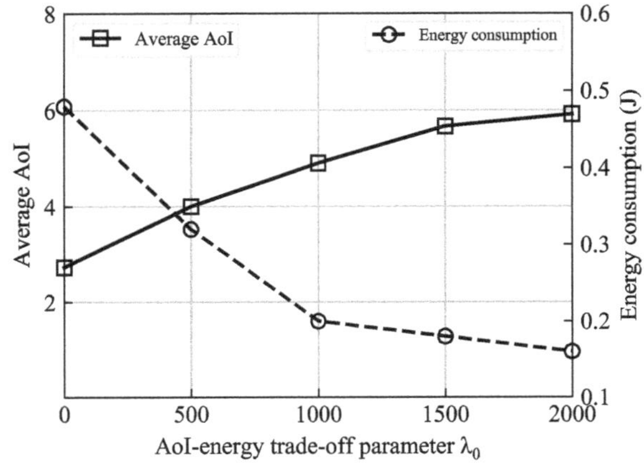

(a) AoI vs. energy consumption

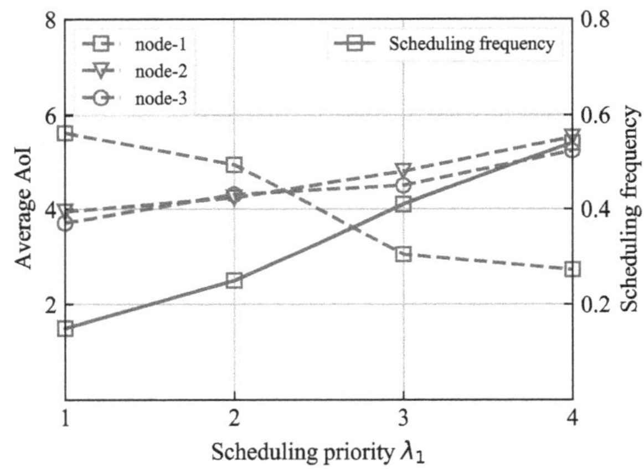

(b) Scheduling frequencies vs. priorities

ability to handle heterogeneous QoS requirements by adjusting the priority weights λ_k. Consider a network with three sensor nodes, where $\lambda_2 = \lambda_3 = 2$ are fixed, and λ_1 is gradually increased from 1 to 4. The scheduling frequency, measured over 30 K time slots, is plotted in Fig. 4.6b, showing a nearly linear increase in node-1's access opportunities with λ_1. Concurrently, node-1's AoI decreases significantly as its priority increases, while the AoI of nodes 2 and 3 increases. This indicates that the learned policy dynamically allocates more transmission resources to high-priority nodes by scheduling them more frequently. The resulting performance shift confirms that Algorithm 3 can adapt to changing priority configurations, enabling flexible, QoS-aware scheduling in heterogeneous sensor networks.

4.5 Discussions and Conclusions

In this chapter, we have studied an IRS-assisted and wireless-powered sensor network, where the goal is to minimize the overall AoI of status updates. The AoI minimization problem is formulated as a mixed-integer non-convex optimization program, accounting for dynamic channel conditions and energy causality constraints. To address the complexity of joint scheduling, energy transfer, and passive beamforming, we propose a hierarchical learning framework that combines model-free DRL in the outer loop with efficient optimization methods in the inner loop. The outer loop learns long-term scheduling and activation policies, while the inner loop computes real-time beamforming and resource allocation solutions. Extensive simulations demonstrate that the proposed algorithm significantly reduces average AoI compared to conventional baselines, while achieving controllable fairness among sensor nodes. In particular, the hierarchical PPO approach achieves higher cumulative rewards and faster convergence than hierarchical DQN and greedy policies. The performance gain is especially prominent in networks with small-size IRSs, where intelligent scheduling and resource coordination are crucial to compensate for limited beamforming gain.

References

1. S. Gong, L. Cui, B. Gu, B. Lyu, D.T. Hoang, D. Niyato, Hierarchical deep reinforcement learning for age-of-information minimization in IRS-aided and wireless-powered wireless networks. IEEE Trans. Wireless Commun. **22**(11), 8114–8127 (2023)
2. M.A. Abd-Elmagid, N. Pappas, H.S. Dhillon, On the role of age of information in the internet of things. IEEE Commun. Mag. **57**(12), 72–77 (2019)
3. Y. Sun, E. Uysal-Biyikoglu, R.D. Yates, C.E. Koksal, N.B. Shroff, Update or wait: how to keep your data fresh. IEEE Trans. Inf. Theory **63**(11), 7492–7508 (2017)
4. Z. Bao, Y. Dong, Z. Chen, P. Fan, K.B. Letaief, Age-optimal service and decision processes in internet of things. IEEE Internet Things J. **8**(4), 2826–2841 (2021)
5. C. Xu, Y. Xie, X. Wang, H.H. Yang, D. Niyato, T.Q.S. Quek, Optimal status update for caching enabled IoT networks: a dueling deep r-network approach. IEEE Trans. Wirel. Commun. **20**(12), 8438–8454 (2021)
6. C. Zhou, G. Li, J. Li, Q. Zhou, B. Guo, FAS-DQN: freshness-aware scheduling via reinforcement learning for latency-sensitive applications. IEEE Trans. Comput. **71**(10), 2381–2394 (2021)
7. E.T. Ceran, D. Gündüz, A. György, A reinforcement learning approach to age of information in multi-user networks with HARQ. IEEE J. Sel. Area. Commun. **39**(5), 1412–1426 (2021)
8. S. Gong, X. Lu, D.T. Hoang, D. Niyato, L. Shu, D.I. Kim, Y.-C. Liang, Toward smart wireless communications via intelligent reflecting surfaces: a contemporary survey. IEEE Commun. Surv. Tutor. **22**(4), 2283–2314 (2020)
9. B.-M. Robaglia, A. Destounis, M. Coupechoux, D. Tsilimantos, Deep reinforcement learning for scheduling uplink IoT traffic with strict deadlines, in *Proceedings of IEEE GLOBECOM*, Dec 2021, pp. 1–6
10. J. Feng, W. Mai, X. Chen, Simultaneous multi-sensor scheduling based on double deep Q-learning under multi-constraint, in *Proceedings of IEEE ICCC*, Jul 2021, pp. 224–229

11. Y. Gu, H. Chen, Y. Zhou, Y. Li, B. Vucetic, Timely status update in internet of things monitoring systems: an age-energy tradeoff. IEEE Internet Things J. **6**(3), 5324–5335 (2019)
12. X. Wu, X. Li, J. Li, P. C. Ching, H.V. Poor, Deep reinforcement learning for IoT networks: age of information and energy cost tradeoff, in *Proceedings of IEEE GLOBECOM*, Dec 2020, pp. 1–6
13. M. Hatami, M. Jahandideh, M. Leinonen, M. Codreanu, Age-aware status update control for energy harvesting IoT sensors via reinforcement learning, in *Proceedings of IEEE PIMRC*, Oct 2020, pp. 1–6
14. S. Leng, A. Yener, Age of information minimization for an energy harvesting cognitive radio. IEEE Trans. Cogn. Commun. Network. **5**(2), 427–439 (2019)
15. A. Arafa, J. Yang, S. Ulukus, H.V. Poor, Age-minimal transmission for energy harvesting sensors with finite batteries: online policies. IEEE Trans. Inf. Theory **66**(1), 534–556 (2020)
16. X. Ling, J. Gong, R. Li, S. Yu, Q. Ma, X. Chen, Dynamic age minimization with real-time information preprocessing for edge-assisted IoT devices with energy harvesting. IEEE Trans. Netw. Sci. Eng. **8**(3), 2288–2300 (2021)
17. M. Hatami, M. Leinonen, M. Codreanu, Aoi minimization in status update control with energy harvesting sensors. IEEE Trans. Commun. **69**(12), 8335–8351 (2021)
18. M.A. Abd-Elmagid, H.S. Dhillon, N. Pappas, A reinforcement learning framework for optimizing age of information in RF-powered communication systems. IEEE Trans. Commun. **68**(8), 4747–4760 (2020)
19. A.H. Zarif, P. Azmi, N. Mokari, M.R. Javan, E. Jorswieck, AoI minimization in energy harvesting and spectrum sharing enabled 6G networks. IEEE Trans. Green Commun. Network. **6**(4), 2043–2054 (2022)
20. S. Leng, A. Yener, Learning to transmit fresh information in energy harvesting networks. IEEE Trans. Green Commun. Network. **6**(4), 2032–2042 (2022)
21. L. Liu, K. Xiong, J. Cao, Y. Lu, P. Fan, K.B. Letaief, Average AoI minimization in UAV-assisted data collection with RF wireless power transfer: a deep reinforcement learning scheme. IEEE Internet Things J. **9**(7), 5216–5228 (2022)
22. T. Bai, C. Pan, Y. Deng, M. Elkashlan, A. Nallanathan, L. Hanzo, Latency minimization for intelligent reflecting surface aided mobile edge computing. IEEE J. Sel. Area. Commun. **38**(11), 2666–2682 (2020)
23. Y. Cao, T. Lv, Z. Lin, W. Ni, Delay-constrained joint power control, user detection and passive beamforming in intelligent reflecting surface-assisted uplink mmwave system. IEEE Trans. Cogn. Commun. Netw. **7**(2), 482–495 (2021)
24. A. Muhammad, M. Elhattab, M.A. Arfaoui, A. Al-Hilo, C. Assi, Age of information optimization in RIS-assisted wireless networks. IEEE Trans. Netw. Serv. Manag. **21**(1), 925–938 (2024)
25. M. Samir, M. Elhattab, C. Assi, S. Sharafeddine, A. Ghrayeb, Optimizing age of information through aerial reconfigurable intelligent surfaces: a deep reinforcement learning approach. IEEE Trans. Veh. Technol. **70**(4), 3978–3983 (2021)
26. G. Cuozzo, C. Buratti, R. Verdone, A 2.4-GHz LoRa-based protocol for communication and energy harvesting on industry machines. IEEE Internet Things J. **9**(10), 7853–7865 (2022)
27. J. Yao, N. Ansari, Wireless power and energy harvesting control in IoD by deep reinforcement learning. IEEE Trans. Green Commun. Netw. **5**(2), 980–989 (2021)
28. B. Lyu, P. Ramezani, D.T. Hoang, S. Gong, Z. Yang, A. Jamalipour, Optimized energy and information relaying in self-sustainable IRS-empowered WPCN. IEEE Trans. Commun. **69**(1), 619–633 (2021)
29. B. Lyu, D.T. Hoang, S. Gong, D. Niyato, D.I. Kim, IRS-based wireless jamming attacks: when jammers can attack without power. IEEE Wirel. Commun. Lett. **9**(10), 1663–1667 (2020)
30. S. Boyd, S.P. Boyd, L. Vandenberghe, *Convex Optimization* (Cambridge University Press, Cambridge, 2004)
31. Y. Zou, Y. Long, S. Gong, D. T. Hoang, W. Liu, W. Cheng, D. Niyato, Robust beamforming optimization for self-sustainable intelligent reflecting surface assisted wireless networks. IEEE Trans. Cogn. Commun. Netw. 1 (2021)

32. Y. Zou, S. Gong, J. Xu, W. Cheng, D.T. Hoang, D. Niyato, Wireless powered intelligent reflecting surfaces for enhancing wireless communications. IEEE Trans. Veh. Technol. **69**(10), 12369–12373 (2020)
33. J. Schulman, F. Wolski, P. Dhariwal, A. Radford, O. Klimov, Proximal policy optimization algorithms. arXiv:1707.06347 (2017)
34. T. Jiang, H.V. Cheng, W. Yu, Learning to reflect and to beamform for intelligent reflecting surface with implicit channel estimation. IEEE J. Sel. Area. Commun. **39**(7), 1931–1945 (2021)
35. I. Kadota, A. Sinha, E. Uysal-Biyikoglu, R. Singh, E. Modiano, Scheduling policies for minimizing age of information in broadcast wireless networks. IEEE/ACM Trans. Netw. **26**(6), 2637–2650 (2018)

Chapter 5
Hierarchical MADRL for Mobile Edge Computing

5.1 Introduction

Wireless networks are inherently multi-agent systems with heterogeneous devices and diverse computational capabilities, including the terrestrial and aerial users, edge devices, and the remote servers. A hierarchical decomposition is indispensable for a large-scale wireless network to be fast converging and reliable in autonomous decision making. One typical example is the decentralized MEC system with multiple edge servers providing computational services to wireless edge users. Each edge server manages its computational resource allocation, power control, beamforming, and scheduling policies for the edge users. While each edge user needs to decide the offloading and transmission strategies based on its estimate of the traffic demand and local resource supply. Another example is the UAV-assisted wireless network serving a large set of GUs distributed over a large geographical area. Multiple UAVs can collaboratively optimize individuals' trajectories and service regions. Each UAV can be viewed as an independent agent making its own decisions (e.g., beamforming, scheduling, and trajectory planning strategies) based on local observations to serve the GUs under its coverage. These examples highlight a common heterogeneous and decentralized network architecture. This motivates a hierarchical multi-agent learning framework to decompose the network optimization problem into several subproblems that can be individually solved at different network layers. This hierarchical framework features a high-level learning loop for network controllers or access points, e.g., the edge servers and UAVs, and the low-level learning loop for edge users. Each layer only focuses on a distinct part of the control variables with reduced dimensionality and complexity.

Our case study in this chapter focuses on the heterogeneous decomposition of edge severs and end users in the MEC system. Moreover, we deploy MADRL at low-level edge users to coordinate the end users' offloading and transmission decisions. The implementation of MADRL usually demands for a large set of training data from all users. This makes it impractical for implementation in resource-limited wireless

S. Gong et al., *Optimization-Driven Deep Reinforcement Learning for Wireless Networks*, Wireless Networks, https://doi.org/10.1007/978-3-032-22997-7_5

networks. MADRL's centralized training requires all users' states and actions. This implies a rapidly increasing dimensionality and thus results in a slow training performance and poor scalability. In this chapter, we improve the learning performance of the low-level MADRL by deploying more efficient model-based optimization at the high-level agent. The following analysis and algorithm design have been presented in our earlier works [1, 2]. This chapter refines and reorganizes that content to ensure consistency, clarity, and coherence within the book's scope.

5.2 Case Study in Hybrid MEC Offloading

The future IoT aims at connecting billions of wireless devices with limited computation capability and energy supply, such as healthcare wearable devices and industrial wireless sensors [3]. The explosive IoT interconnections enable the emerging of new demanding applications, such as augmented reality, autonomous driving, and interactive gaming, which are highly computation-intensive and latency-sensitive. Moreover, the development of IoT applications will be no longer limited to terrestrial networks, but expanding into the future space-air-ground integrated system [4]. MEC is a key technology for enabling these demanding applications and alleviating the need for intensive computation at such IoT devices, e.g., [5, 6]. The key idea of MEC is to deploy resource-rich MEC servers in close proximity to the resource-constrained IoT devices. As such, the MEC servers' computation capabilities can be shared among all end devices by computation offloading and optimal resource allocation, significantly enhancing the overall energy efficiency and reducing processing latency.

5.2.1 Motivations and Contributions

Conventional data offloading via RF communications is inherently power demanding. This high energy cost may prevent the energy-constrained IoT devices from using the MEC service. Thus, a more energy-efficient offloading strategy is required for energy constrained IoT to fully exploit MEC's potential. Without using power-consuming devices, the passive wireless backscatter communications feature with extremely low power consumption, e.g., [7, 8]. This also comes with the cost of a low data rate and increased vulnerability to channel fluctuations. Such complementary strengths and weaknesses in transmission capabilities and energy consumption motivate us to exploit a hybrid approach for the integration of active RF communications and backscatter communications to improve the overall MEC offloading performance, e.g., [9, 10]. This case study aims at investigating an energy-efficient hybrid MEC offloading strategy in such a hybrid radio network. Here, each edge user can dynamically choose its offloading mode, i.e., either the active RF communications or the backscatter communications, depending on its energy supply, workload size, channel conditions, and delay tolerance. Our objective is to maximize the system's energy efficiency while guaranteeing the completion of all users' computational workloads. This is achieved by jointly optimizing the users' transmission scheduling and workload allocation among local computation, active RF offloading, and passive backscactter offloading.

The capability of switching between active and passive offloading modes offers resource-constrained edge users greater flexibility in balancing energy consumption between communication and computation. However, this flexibility also significantly increases the complexity of making optimal offloading decisions. The first major challenge stems from uncertainties in users' workload and energy availability under dynamic network environments. For instance, the volume of user's traffic demand may fluctuate considerably with the user's mobility, while the energy consumption for data offloading is heavily influenced by time-varying channel conditions. The second challenge arises from intense competition among users for limited computational resources at MEC servers, which are typically co-located with BSs [11]. This necessitates joint optimization of computation resource allocation and transmission control, such as power allocation and transmit beamforming. This introduces tight coupling among different users due to the broadcast nature of wireless channels. Consequently, performance optimization in MEC systems generally amounts to solving a high-dimensional control problem that is particularly challenging in dynamic network environments with incomplete or uncertain state information. Conventional optimization often fails to adapt efficiently under such conditions, calling for more intelligent and scalable solutions.

DRL has emerged as a more flexible and robust approach for adapting MEC offloading decisions through continuous interaction with uncertain and dynamic network environments [12]. It allows each edge user to make localized offloading decisions, such as transmission scheduling and workload allocation, based on its own observations, including channel conditions, computational workload, and energy availability. In this case study, we propose a hierarchical MADRL framework to enhance learning efficiency and scalability. The maximization of system energy efficiency necessitates the joint optimization of the AP's beamforming strategy and user-level decisions on energy and workload allocation among local computing, passive offloading, and active offloading. Given the stochastic nature of workload arrivals and channel conditions, we formulate the energy-efficiency maximization problem into MDP and propose the hierarchical MADRL solution. Instead of relying on a single centralized agent at the AP, the hierarchical MADRL decentralizes decision-making at both the AP and edge users, by introducing a high-level agent at the AP and multiple low-level agents at individual edge users. The high-level agent coordinates global strategies such as beamforming and resource allocation, while low-level agents optimize local offloading and computation policies. Such a hierarchical decomposition improves convergence, scalability, and adaptability in complex and dynamic network environments. The main technical contributions of this case study are summarized as follows:

- Hybrid offloading for improved energy efficiency: We propose a multi-user hybrid offloading system to balance edge devices' energy consumption between local computation and data offloading. By enabling dynamic switching between active and passive offloading modes, the system provides greater flexibility in resource management to improve energy efficiency. For example, energy-constrained users may employ passive offloading to reduce power consumption, while users requiring higher transmission rates can leverage active offloading for improved performance, thereby enhancing overall energy efficiency.
- Hierarchical MADRL framework: A hierarchical MADRL framework is devised to maximize energy efficiency while reducing computational complexity and communication overhead. In this framework, a high-level agent optimizes the AP's beamforming strategy while low-level agents deployed at user devices dynamically allocate time and workload among local computation, passive offloading, and active offloading modes. The user agents' action further guides the AP to refine the high-level agent's beamforming policy, establishing a closed-loop interaction between the AP and users to enhance system-wide performance continuously.
- Accelerating user agents' MADRL: To improve learning efficiency for the low-level user agents, we introduce the optimization-driven learning algorithm that integrates model-based optimization into the MADRL framework. The embedded optimization module efficiently generates action estimates by solving computationally efficient approximations of the local offloading subproblems, such as workload partitioning and time allocation. These action estimates serve as informed priors that effectively guide MADRL's learning process. In particular, each edge user independently updates its offloading policy using a model-free DRL algorithm with significantly reduced state and action spaces. This can accelerate convergence and improve sample efficiency. Moreover, the decentralized policy updates at individual edge users enable a parallelized training across users, enhancing scalability and reducing communication overhead.

5.2.2 *Related Works*

Recently, model-free DRL methods have been largely and successfully applied to MEC control problems [13]. By interacting with dynamic network environments, these methods enable users to autonomously adapt their offloading strategies in response to fluctuating workload demands and time-varying channel conditions. In multi-server MEC environments, various DRL-based approaches have been proposed. The work in [14] uses DQN to optimize MEC server selection and resource allocation that maximize energy efficiency while meeting task deadline constraints. Similarly, the authors in [15] applied DQN to optimize the users' server selection and offloading rates based on battery levels, historical transmission rates, and energy harvesting capabilities. In [16], DQN was employed to learn offloading policies that maximize a multi-factor utility function incorporating user satisfaction with execution delay, task drops, queuing delay, outage events, and MEC service costs. DQN was also adopted in a vehicular edge computing network to adapt the charging scheduling policy for low-battery electric vehicles [17], aiming to minimize overall overhead under fluctuating electricity prices and charging station availability. Beyond value-based DRL methods, policy-based DRL algorithms have also shown strong performance in MEC offloading problems. The authors of [18] introduced an actor-critic DRL framework to minimize end-to-end delay through joint optimization of content caching, computation offloading, and radio resource allocation.

The MADRL framework has been increasingly adopted in the literature to optimize multi-user MEC offloading strategies. A straightforward approach is independent DQN, where each user operates as a separate DQN agent that makes decisions based solely on its local observations [19]. The asynchronous advantage actor-critic (A3C) algorithm, employing a parallel training architecture, is another prominent approach introduced in [20]. Multiple agents are allowed to interact with their environments simultaneously and update a shared actor-critic network asynchronously using local experiences. Each agent individually performs training updates by using the conventional actor-critic methods such as DDPG. As such, A3C essentially represents a parallelized implementation of the single-agent DDPG algorithm, enhancing sample efficiency and exploration through decentralized execution. To better capture agents' interactions, the DDPG algorithm has been extended to the multi-agent setting through the MADDPG framework, which adopts a centralized training with decentralized execution (CTDE) paradigm [21]. This framework has been effectively applied in various MEC scenarios. For example, the authors in [22] applied MADDPG to minimize energy consumption under latency constraints in small cell networks by jointly optimizing computation offloading and interference coordination. Similarly, the authors in [23] utilized MADDPG in multi-UAV-assisted MEC systems for joint trajectory design and computation offloading. Further building on the CTDE principle, the authors in [24] proposed a distributed MADRL framework featuring a centralized critic and multiple decentralized actors to minimize average task completion time in a multi-server MEC system with task migration. This distributed framework enables global reward evaluation while preserving local decision-making autonomy, achieving a balance between performance and scalability in complex and dynamic MEC systems.

MADRL algorithms often demand extensive offline training and may suffer from instability, high sample complexity, and reduced flexibility as the number of user agents grows. These challenges stem from the exponential growth of the joint state and action spaces, as well as the non-stationarity due to multiple agents' concurrent learning. Several strategies have been proposed to address these limitations and enhance learning efficiency under these constraints. A prominent approach involves decomposing the overall optimization problem into tractable subproblems, each solved with a suitable method tailored to its structure. For instance, the authors in [25] decoupled the system design into user association and computation offloading, solving the former using a two-sided matching algorithm and the latter via a DRL-based approach. Similarly, the authors in [26] applied a hybrid framework where a multi-DQN architecture is employed to learn energy-efficient sleep-scheduling policies for edge servers, while an iterative optimization algorithm handles resource allocation. Such decomposition not only reduces the dimensionality of the control problem but also enables faster convergence by combining model-based optimization with model-free learning.

Another line of research accelerates learning by incorporating expert knowledge or model-based optimization. These methods aim to guide exploration and reduce reliance on random trial-and-error, thereby improving convergence speed and sample efficiency. For instance, the authors in [27] addressed the optimal deployment of multiple UAVs for differentiated service provisioning to GUs. A multi-agent learning method was developed to maximize profits for both UAVs and users in a decentralized manner by leveraging expert demonstrations through imitation learning. By learning from high-quality expert policies, agents can bypass inefficient exploration and quickly adopt effective strategies. Furthermore, model-based optimization can enhance learning performance in different ways. Optimization techniques can generate large sets of labeled training samples for offline pre-training. For example, the authors in [28] alternately applied block coordinate descent (BCD) and SCA to iteratively optimize computation offloading, bit allocation, and UAV's trajectory planning strategies. The solutions from these optimization methods serve as supervisory signals to initialize or regularize the DRL agent's policy, effectively bootstrapping the learning process. Beyond offline initialization, model-based optimization can also guide online learning. The work in [29] proposed a sampling scheme to generate multiple candidate actions, each evaluated by an optimization module to estimate its potential performance, allowing the agent to select near-optimal actions during exploration. Similarly, the authors in [30] integrated a model-based optimization module directly into the DRL's critic. This enables the critic to assess each action and chooses a more preferable action based on analytical performance evaluation rather than random exploration. By embedding domain knowledge or model information into learning, this hybrid approach achieves faster convergence, improved stability, and higher reward performance.

5.3 System Model and Problem Formulation

In the case study, we consider a wireless network comprising a multi-antenna AP equipped with M antennas and serving N wireless-powered edge users, indexed by the set $\mathcal{N} = \{1, 2, \ldots, N\}$. Following the framework in [31], each edge user is capable of harvesting RF energy from the AP's beamforming signals to replenish its battery and support its operational tasks, such as data transmission or local computation. Owing to limited energy and computational resources at individual edge users, each user may offload a portion of its data or computational workload to a nearby MEC server, which is co-located with the AP. These edge users may be deployed as wireless-powered wearable devices for real-time health monitoring and analysis of healthcare information. The collected sensing data can subsequently be processed using ML algorithms to enable classification, prediction, and informed decision-making. The proposed system model is readily extensible to scenarios involving multiple MEC servers or APs. The complex channel vector between the AP and the i-th edge user is denoted by $\mathbf{h}_i \in \mathbb{C}^{M \times 1}$, while the complex channel coefficient between the i-th and j-th users is represented by $g_{i,j} \in \mathbb{C}$. Based on its energy state, channel conditions, and delay tolerance, each user dynamically determines an optimal workload partitioning strategy between local computation and computation offloading. Specifically, the computational workload of each user is partitioned into different components: (i) a portion executed locally, (ii) a portion offloaded to the MEC server via either active RF communication or (iii) passive backscatter communication.

5.3.1 Workload and Offloading Time Allocation

We consider a time-division protocol that enables multiple edge users to offload their data or computational workload to a MEC server. The offloading time frame, denoted by T_o, is partitioned into multiple non-overlapping time slots, with each edge user $i \in \mathcal{N}$ assigned a dedicated time slot T_i. Each user may choose to process its workload either locally or remotely via offloading, depending on its available energy, instantaneous channel conditions, and allocated time slot. Let ℓ_i represent the total workload of the i-th user. We assume that the MEC server can return the processed results instantaneously, a reasonable assumption justified by the typically negligible size of the output data relative to the input [31]. The processed results from both the local computation and the MEC server can be aggregated to support the user's subsequent decision-making tasks. To ensure timely execution, the entire workload of each user must be fully processed either locally or at the MEC, before the expiration of its allocated time slot T_i. Consequently, the workload allocation for each user must satisfy the following completion constraint:

$$\ell_{a,i} + \ell_{b,i} + \ell_{c,i} \geq \ell_i, \quad \forall i \in \mathcal{N}, \tag{5.1}$$

where $\ell_{a,i}$ and $\ell_{b,i}$ denote the portions of the offloaded workload transmitted by user i to the MEC server via active RF and passive communication modes, respectively. The third component, $\ell_{c,i}$, represents the workload processed locally by user i. The energy consumption associated with local computation depends on the user's computational capability and the size of the locally executed workload $\ell_{c,i}$, for all $i \in \mathcal{N}$. Let $f_{c,i}$ denote the operating frequency of the local processor, and let $\phi_{c,i}$ represent the number of processing cycles required to process one unit of workload. Given the local workload size $\ell_{c,i}$, the corresponding local computation time is expressed as $t_{c,i} = \ell_{c,i}/r_{c,i} = \ell_{c,i}\phi_{c,i}/f_{c,i}$, where $r_{c,i} \triangleq f_{c,i}/\phi_{c,i}$ is defined as the local processing rate of user i. Furthermore, the energy consumed per processing cycle is modeled as $k_e f_{c,i}^2$, where k_e captures the energy efficiency of the local processor [32]. Consequently, the energy consumption for local computation is given by:

$$e_{c,i} = t_{c,i} k_e f_{c,i}^3 = \phi_c k_e f_{c,i}^2 \ell_{c,i}. \tag{5.2}$$

The limited energy budgets and constrained computational resources at individual edge users motivate the offloading of workloads to the MEC server, which can be accomplished through either passive backscatter or active RF communication. To accommodate both offloading modes within the allocated time slot, we further partition each user's time slot T_i into two distinct sub-slots, i.e., an active offloading sub-slot $t_{a,i}$ and a passive offloading sub-slot $t_{b,i}$. Consequently, the total workload of each user must be fully processed either locally or via offloading within its assigned time slot T_i, imposing the following time constraint:

$$t_{a,i} + t_{b,i} \leq T_i \text{ and } t_{c,i} \leq T_i, \quad \forall i \in \mathcal{N}. \tag{5.3}$$

Note that local computation can be performed concurrently with data offloading. Therefore, the local processing time must satisfy $t_{c,i} \leq T_i$, where T_i is the total time budget for user i. Let $r_{a,i}$ and $r_{b,i}$ denote the data offloading rates for active and passive communication modes, respectively. Thus, the workload completion constraint in (5.1) can be equivalently expressed as

$$t_{a,i} r_{a,i} + t_{b,i} r_{b,i} + \ell_{c,i} \geq \ell_i, \quad \forall i \in \mathcal{N}. \tag{5.4}$$

This implies that the optimal workload or time allocation strategy depends on the user's workload demand, energy supply, and processing rates in different modes. In the sequel, we derive the offloading rates in active and passive modes, respectively.

5.3.2 *Hybrid Active and Passive Offloading*

We allow resource-constrained edge users to split their workload between active and passive offloading modes. Let $\beta_{a,i}$ denote the transmit power of the i-th edge user. The corresponding received signal at the AP is expressed as $\mathbf{y}_{a,i} = \sqrt{\beta_{a,i}} \mathbf{h}_i x_i(t) + \mathbf{v}_o$, where $x_i(t)$ represents the unit-power information signal transmitted by user i, $\mathbf{h}_i$ denotes the channel vector between user i and the AP, and $\mathbf{v}_o$ is the additive white Gaussian noise vector at the multi-antenna AP. Assuming the AP employs optimal MRC for signal detection, the normalized active offloading rate is given by:

$$r_{a,i} = \log_2 \left(1 + \beta_{a,i} |\mathbf{h}_i|^2\right), \tag{5.5}$$

where we assume unit noise power for simplicity. Given the active offloading rate $r_{a,i}$, the required transmit power for active RF communication is determined by $\beta(r_{a,i}) \triangleq (2^{r_{a,i}} - 1) / |\mathbf{h}_i|^2$. Consequently, the total power consumption associated with active offloading is given by $p_{a,i} = \beta\left(r_{a,i}\right) + p_o$, which comprises the transmit power $\beta(r_{a,i})$ and a fixed circuit power consumption p_o necessary to operate the RF front-end. It is evident that the active offloading rate $r_{a,i}$ is inherently dependent on the channel realization $\mathbf{h}_i$, and thus becomes a stochastic quantity in dynamic wireless environments.

In passive offloading, the AP emits carrier signals, while the edge user concurrently backscatters its workload information by modulating the incident RF signals through reflection. Let $\mathbf{w}_i \in \mathbb{C}^{M\times 1}$ denote the normalized beamforming vector associated with the i-th edge user. The AP's RF beamforming signal during the i-th sub-slot is expressed as $\mathbf{u}_i(t) = \sqrt{p_A}\,\mathbf{w}_i s$, where p_A represents the AP's constant transmit power and $s \in \mathbb{C}$ denotes a unit-power random information symbol. Consequently, the incident RF signal at the i-th edge user is given by $c_i(t) = \mathbf{h}_i^H \mathbf{u}_i(t)$. Meanwhile, the i-th edge user modulates its information signal $x_i(t)$ onto the incident RF signal $c_i(t)$ by adjusting its reflection coefficient according to $\Gamma(t) = \Gamma_o x_i(t)$, where Γ_o denotes an antenna-specific constant coefficient [33]. The resulting signal received at the AP is expressed as:

$$\mathbf{y}_{b,i}(t) = \Gamma_o \mathbf{h}_i^H x_i(t) c_i(t) + \mathbf{F}^H \mathbf{u}_i + \mathbf{v}_o, \tag{5.6}$$

where $\mathbf{F}$ represents the loop-back self-interference channel matrix and $\mathbf{v}_o$ denotes the additive noise. In (5.6), the first term carries the desired workload information $x_i(t)$, whereas the second term $\mathbf{F}^H \mathbf{u}_i$ corresponds to self-interference induced by the AP's concurrent transmission. Under the assumption of ideal full-duplex operation and perfect self-interference cancellation, this interference component can be effectively suppressed. By employing MRC, the effective received signal power at the AP is given by $p_A|\Gamma_o|^2\|\mathbf{h}_i\|^2|\mathbf{h}_i^H\mathbf{w}_i|^2$. Consequently, the normalized achievable data rate for passive offloading is approximated as follows:

$$r_{b,i} = \log_2\left(1 + \frac{p_A|\Gamma_o|^2\|\mathbf{h}_i\|^2|\mathbf{h}_i^H\mathbf{w}_i|^2}{|\mathbf{h}_i^H\mathbf{v}_o|^2}\right). \tag{5.7}$$

In practical deployments, residual self-interference may persist due to imperfect cancellation, and the aggregate disturbance $\mathbf{F}^H\mathbf{u}_i + \mathbf{v}_o$ complicates an exact characterization of the passive offloading rate. Nevertheless, when the power of the information-bearing signal (i.e., the first term in (5.6)) dominates over that of the residual interference and noise, the expression in (5.7) serves as a valid approximation of the achievable rate.

5.3.3 Energy Minimization For Hybrid MEC Offloading

The total energy consumption of the MEC system comprises two primary components: the energy consumed by the AP for RF beamforming and that consumed by the MEC server for workload reception and processing. Given the AP's beamforming signal $\mathbf{u}_i(t) = \sqrt{p_A}\,\mathbf{w}_i s$ during the i-th time slot, the associated energy consumption is expressed as $e_{b,i} = p_A t_{b,i}$, where $t_{b,i}$ denotes the time allocated to passive offloading. The MEC server incurs energy costs primarily during data reception and computational execution. Let $p_{a,r}$ and $p_{b,r}$ represent the receiver power consumption under active and passive offloading modes, respectively, with the typical relation $p_{a,r} > p_{b,r}$. For user i, the energy consumed in data reception is given by $e_{r,i} \triangleq p_{a,r}t_{a,i} + p_{b,r}t_{b,i}$, where $t_{a,i}$ and $t_{b,i}$ are the time durations allocated to active and passive offloading, respectively. The remaining energy expenditure arises from workload computation, which scales with the total amount of workload offloaded to the MEC server, denoted by $\ell_{o,i} \triangleq \ell_{a,i} + \ell_{b,i}$. Analogous to the local computation energy model in (5.2), the MEC server's computational energy consumption is modeled as $e_{o,i} = \phi_o k_o f_o^2 \ell_{o,i}$, where k_o represents the number of processing cycles required per unit workload, f_o denotes the server's processor frequency, and ϕ_o is the energy efficiency coefficient of the MEC server's processor. Consequently, the overall energy consumption of the MEC system for user i is specified as:

$$E_o = \sum_{i \in \mathcal{N}} \Big(p_A t_{b,i} + p_{a,r}t_{a,i} + p_{b,r}t_{b,i} + \phi_o k_o f_o^2 \ell_{o,i} \Big). \tag{5.8}$$

Active offloading relies on the energy stored in the user's battery. Given the AP's beamforming signal $\mathbf{u}_i(t) = \sqrt{p_A}\,\mathbf{w}_i s$, the RF power received by user j during the i-th time slot can be approximated as follows:

$$p_{j,i}^h \triangleq \mathbb{E}[|\mathbf{h}_j^H \mathbf{x}_i(t)|^2] = p_A |\mathbf{h}_j^H \mathbf{w}_i|^2,$$

where $\mathbf{h}_j \in \mathbb{C}^{M\times 1}$ denotes the channel vector from the AP to user j. This received RF power serves as the input to a nonlinear energy harvesting model, enabling estimation of the harvested energy E_j^h, as detailed in [10, 31]. The edge users' self-stainability requires that the overall energy consumption is limited as follows:

$$t_{a,j}p_{a,j} + e_{c,j} \le \min\left\{E_j + E_j^h, E_{\max,j}\right\}, \quad \forall j \in \mathcal{N}, \tag{5.9}$$

where E_j denotes the initial energy level in battery of user j, and $E_{\max,j}$ presents its maximum battery capacity. The energy consumption for local computation $e_{c,j}$ is given by (5.2). The user's energy consumption associated with passive offloading is omitted in our formulation, owing to the extremely low power requirements inherent to backscatter communications. Specifically, the power consumption of backscatter transmission is typically below 1 μW, whereas active RF communication consumes on the order of hundreds of milliwatts [8]. Moreover, the power consumption in backscatter communications can be treated as a constant that does not depend on the system variables involved in the optimization. Consequently, it has no bearing on the optimal solution and is thus excluded from the energy model. As indicated by the energy budget constraint in (5.9), the user's workload allocation is fundamentally limited by its available energy, which is implicitly coupled with the time allocations assigned to other users due to the shared nature of the wireless medium.

We aim to minimize the total energy consumption required to complete all users' workloads by jointly optimizing the AP's beamforming vectors $\mathbf{w}_i$ and the edge users' hybrid offloading decisions. Specifically, for each user $i \in \mathcal{N}$, the offloading decisions include time allocation $\boldsymbol{t}_i \triangleq (t_{a,i}, t_{b,i}, t_{c,i})$ and workload division $\boldsymbol{\ell}_i \triangleq (\ell_{a,i}, \ell_{b,i}, \ell_{c,i})$, formulated as follows:

$$\min_{\mathbf{w}_i, \boldsymbol{t}_i, \boldsymbol{\ell}_i} \sum_{i \in \mathcal{N}} \Big(e_{b,i} + e_{r,i} + e_{o,i} \Big), \quad \text{s.t. (5.3); (5.4); and (5.9).} \tag{5.10}$$

Notably, the energy available to each edge user is harvested exclusively from the AP's beamforming signals. Consequently, minimizing the system-wide energy consumption is equivalent to reducing the aggregate energy expenditure of the AP and the MEC server. The objective function in problem (5.10) is linear with respect to the time allocation variables $\boldsymbol{t}_i$. Constraints (5.3) and (5.4) respectively enforce temporal feasibility and workload consistency for each edge user, while the energy budget constraint (5.9) introduces inter-user coupling through the joint dependence on $\boldsymbol{\ell}_i$ and $\boldsymbol{t}_i$.

The AP's beamforming vector $\mathbf{w}_i$ simultaneously governs both the offloading rates $r_{a,i}$ and $r_{b,i}$, as defined in (5.4), and the amount of wireless power transferred to the edge users. In particular, the passive offloading rate $r_{b,i}$, as expressed in (5.7), is a function of $\mathbf{w}_i$ during the time slot T_i. Moreover, any user $j \neq i$ can harvest RF energy from the AP's beamforming signals intended for other users, which in turn powers its active offloading and local computation tasks. This mutual dependence creates strong inter-user couplings, rendering the independent optimization of individual offloading strategies highly challenging. An additional source of complexity in solving (5.10) stems from the uncertain and dynamic nature of the wireless environment. Time-varying channel conditions lead to fluctuating offloading rates, and inaccurate workload estimation may result in workload outage, i.e., the inability to successfully process a user's workload within the allocated time slot. These challenges motivate the development of a robust learning framework capable of adaptively coordinating the AP's beamforming design and users' offloading decisions in response to real-time network dynamics.

5.3.4 Model-Free DRL to Hybrid MEC Offloading

A straightforward approach to implementing DRL is to deploy a centralized agent at either the AP or the MEC server, both of which possess sufficient computational resources and can aggregate global system information from all users to assist decision-making. Such a centralized architecture enables coordinated adaptation of actions for both the AP and individual edge users. In alignment with the design objective of problem (5.10), the action vector $\boldsymbol{a}_t$ at time slot t encompasses all control variables in (5.10), namely the beamforming vectors $\mathbf{w}_i$, time allocations $\boldsymbol{t}_i$, and workload divisions $\boldsymbol{\ell}_i$ for all $i \in \mathcal{N}$. Correspondingly, the reward function $v_t(\boldsymbol{s}_t, \boldsymbol{a}_t)$

is designed to be inversely proportional to the total energy consumption. To handle the continuous action space inherent in this optimization problem, we employ off-policy actor-critic methods such as DDPG or PPO to learn an approximate offloading strategy that effectively addresses the MEC offloading problem formulated in (5.10). The system state $\mathbf{s}_t \in \mathcal{S}$ at each decision epoch aggregates observations from the AP and all edge users in the network. Specifically, for each user i, the local observation $\mathbf{o}_i$ comprises the workload demand ℓ_i, the allocated time duration T_i, and the current energy status E_i, i.e., $\mathbf{o}_i \triangleq (\ell_i, T_i, E_i)$. The AP's observation includes the channel information $(\mathbf{g}_i, \mathbf{h}_i)$, which can be estimated at the AP during a training phase at the beginning of each time slot. Consequently, the global system state is constructed as $\mathbf{s}_t \triangleq (\mathbf{o}_i, \mathbf{g}_i, \mathbf{h}_i)_{i\in\mathcal{N}}$. The action $\mathbf{a}_t \triangleq \left(\mathbf{w}_i, \boldsymbol{t}_i, \boldsymbol{\ell}_i\right)_{i\in\mathcal{N}} \in \mathcal{A}$ encompasses the AP's beamforming vectors $\mathbf{w}_i \in \mathbb{C}^M$ and each user's MEC offloading decisions, including time allocation $\boldsymbol{t}_i$ and workload division $\boldsymbol{\ell}_i$. Upon executing action $\mathbf{a}_t$ in state $\mathbf{s}_t$, the system switches to state $\mathbf{s}_{t+1}$ according to the underlying dynamics.

The reward function $v_t(\mathbf{s}_t, \mathbf{a}_t) \in \mathcal{R}$ is evaluated based on the current state-action pair and is designed to guide the agent toward policies that approximate the optimal solution to the MEC offloading problem [34]. Notably, both computational and energy resources are wasted if a workload outage occurs, i.e., when the system fails to process the entire workload within the allocated time slot. To account for such constraint violations, prior works adopted piecewise reward structures with explicit penalty terms for infeasible actions, particularly those violating delay or workload constraints, e.g., [35]. Following this principle, we design the reward function to be inversely proportional to the total energy consumption and monotonically increasing with respect to the amount of successfully processed workload. Specifically, the reward can be formulated as the product of the successful task completion ratio and the system-wide energy efficiency:

$$v_t(\boldsymbol{s}_t, \boldsymbol{a}_t) = \left(\frac{1}{N}\sum_{i\in\mathcal{N}} x_i\right)^{\sigma} \left(\frac{1}{E_o}\sum_{i\in\mathcal{N}} x_i \ell_i\right), \tag{5.11}$$

where E_o denotes the total energy consumption as defined in (5.8). The binary indicator $x_i \in \{0, 1\}$ reflects whether a workload outage occurs for user i, determined by the feasibility of the constraints in problem (5.10). Specifically, $x_i = 0$ signifies a workload outage, in which case user i's unprocessed workload must be re-executed in a subsequent time slot, leading to additional energy expenditure and inefficient use of computational resources.

The reward in (5.11) consists of two components. The first term $\frac{1}{N}\sum_{i\in\mathcal{N}} x_i$ acts as a reliability-aware penalty factor, representing the fraction of users whose workloads are successfully processed within the current time slot. This term incentivizes the MEC system to satisfy all task deadlines and workload constraints, thereby enhancing overall service reliability. A higher value indicates a lower probability of workload outage and more effective utilization of available resources. The second term $\frac{1}{E_o}\sum_{i\in\mathcal{N}} x_i \ell_i$ quantifies the system's energy efficiency, defined as the ratio of successfully processed workload to total energy consumption. By jointly considering reliability and efficiency, the reward function balances successful task execution against energy usage. The positive constant coefficient $\sigma > 0$ governs the trade-off between these two objectives. A larger σ places greater emphasis on reliability, making the system more sensitive to workload outages and encouraging conservative offloading decisions that prioritize constraint satisfaction over raw energy savings.

5.4 Hierarchical MADRL for Hybrid MEC Offloading

In this section, we propose a hierarchical multi-agent DDPG (H-MADDPG) framework designed to enhance learning performance by leveraging the inherent structural properties of the hybrid MEC offloading problem in (5.10). The core principle of this framework is to decentralize the decision-making process across both the AP and the edge users, instead of relying on a single centralized decision-making entity located at the AP. Specifically, the H-MADDPG framework comprises a high-level agent situated at the AP and multiple low-level agents deployed at individual edge users, as depicted in Fig. 5.1. At each decision epoch, the AP first computes its high-level action, namely, the beamforming strategy $\mathbf{w}_i$ for each user $i \in \mathcal{N}$, by employing the standard DDPG algorithm. Subsequently, each low-level user $i \in \mathcal{N}$ independently optimizes its MEC offloading time and workload allocation strategies $(\boldsymbol{t}_i, \boldsymbol{\ell}_i)$ via the MADDPG algorithm, conditioned on its local energy status E_i, workload demand ℓ_i, and channel conditions $(\mathbf{h}_i, \mathbf{g}_i)$. The collective decisions of the low-level users define the system state observed by the high-level agent, thereby guiding the AP to refine its beamforming strategy in the subsequent decision epoch. This hierarchical architecture effectively reduces the action space of each agent, thereby enhancing learning efficiency and scalability. In the following subsections, we formulate the high-level optimization problem for the AP and the low-level problems for the edge users, respectively.

5.4.1 High-Level Beamforming Adaptation

The high-level agent is hosted by the AP and is responsible for adapting the AP's beamforming strategy. The high-level system state, denoted by $\mathbf{s}^o$, encompasses the channel conditions between the AP and the edge users, each user's workload demand ℓ_i, time slot T_i, and energy status E_i. Low-level users report their local observations to the AP, upon determining their optimal actions via the low-level MADDPG algorithm. The high-level action is defined as $\mathbf{a}^o \triangleq [\mathbf{w}_1, \mathbf{w}_2, \ldots, \mathbf{w}_N]$, where $\mathbf{w}_i$ represents the beamforming vector assigned by the AP to user $i \in \mathcal{N}$. Given the system state $\mathbf{s}^o$, the high-level agent dynamically adjusts $\mathbf{a}^o$ to satisfy each user's offloading rate requirements while adhering to the energy budget constraints. The high-level reward function, denoted by $v_t^o(\mathbf{s}_t^o, \mathbf{a}_t^o)$, quantifies the overall energy efficiency and

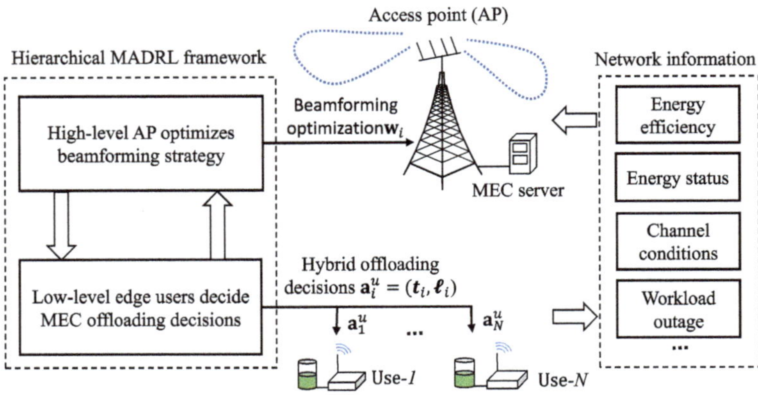

Fig. 5.1 Edge users report local observations and execute actions using MADDPG. The AP adjusts beamforming to meet offloading demands under energy constraints, maximizing a global reward for efficiency and reliability, enabling hierarchical and adaptive coordination between the MEC server and users

system reliability, which can be explicitly aligned with the global design objective of the system, as formalized in (5.11).

5.4.2 Low-Level Time and Workload Allocation

Given the AP's beamforming strategy $\mathbf{a}^o$, each user $i \in \mathcal{N}$ operates as an independent decision-making agent, dynamically adapting its offloading policy to optimize its individual reward based on local observations. The local system state of user i, denoted by $\mathbf{s}_i^u \triangleq (E_i, \ell_i, T_i, \mathbf{h}_i, \mathbf{g}_i)$, comprises its energy status E_i, workload demand ℓ_i, time slot duration T_i, and channel conditions $(\mathbf{h}_i, \mathbf{g}_i)$. The energy budget E_i is harvested from the AP's RF signal via beamforming during time slots $j \neq i$, thereby coupling E_i with the AP's beamforming strategy and the offloading decisions of all other users. The action of user i, denoted $\mathbf{a}_i^u \triangleq (\boldsymbol{t}_i, \boldsymbol{\ell}_i)$, specifies its MEC offloading time allocation and workload distribution across three modes: local computing, active offloading, and passive offloading. User i receives a higher reward when its entire workload is successfully processed within the allocated time slot T_i. However, utilizing the MEC server's computational resources incurs a service cost, which is reflected in the reward design. Accordingly, the reward function for edge user i is defined as follows:

$$v_i^u \triangleq \max\{x_i - \mu_1 \ell_{o,i}/\ell_i - \mu_2 t_{b,i}/T_i, 0\}, \tag{5.12}$$

where the binary variable x_i indicates the absence ($x_i = 1$) or occurrence ($x_i = 0$) of a workload outage. The constants μ_1 and μ_2 represent the unit resource prices associated with the MEC server and the AP, respectively. Specifically, the term $\mu_1 \ell_{o,i}/\ell_i$

captures the normalized cost of offloading a fraction $\ell_{o,i}/\ell_i$ of the workload to the MEC server. Additionally, the AP consumes energy to support passive offloading through RF beamforming, which introduces the second cost term $\mu_2 t_{b,i}/T_i$, proportional to the passive offloading duration $t_{b,i}$ relative to the total time frame T_i.

Given the AP's beamforming strategy, the low-level user agents iteratively refine their offloading time and workload allocation strategies using the conventional MADDPG algorithm. Each user agent's observation of the environment is not only influenced by its own action but also coupled with the actions of all other agents. Consequently, the value function for the i-th user agent is formulated as:

$$V^i_{\pi_i,\boldsymbol{\pi}_{-i}}(\mathbf{S}^u) = \mathbb{E}\left[\sum_{t=0}^{\infty} \gamma^t v_i^u\left(\mathbf{S}^u(t), \mathbf{A}^u(t)\right)\right],$$

where π_i is the policy of the i-th user agent, and $\boldsymbol{\pi}_{-i} \triangleq \{\pi_j\}_{j \neq i}$ denotes the collection of policies of all other user agents. The joint state and action spaces of all user agents are respectively defined as $\mathbf{S}^u = [\mathbf{s}_1^u, \mathbf{s}_2^u, \ldots, \mathbf{s}_N^u]$ and $\mathbf{A}^u = [\mathbf{a}_1^u, \mathbf{a}_2^u, \ldots, \mathbf{a}_N^u]$. Each user's action is determined by its individual policy, i.e., $\mathbf{a}_i^u(t) = \pi_i(\mathbf{s}_i^u(t))$. The interactions among low-level user agents are handled via a centralized training and decentralized execution framework [21]. During the offline training phase, the network controller such as an AP or a MEC server aggregates global information from all users and simultaneously trains each user's critic and actor networks in a centralized fashion. Upon convergence, these well-trained networks are distributed to their respective users for online inference in a decentralized manner. Specifically, each user independently determines its offloading decision using its local actor network. Each user agent's Q-value function, denoted as $Q_i(\mathbf{S}^u, \mathbf{A}^u \mid \boldsymbol{\omega}_i)$, is trained following an approach analogous to that of single-agent DDPG, with the key distinction that Q_i depends on the global state-action tuple $(\mathbf{S}^u, \mathbf{A}^u)$. The target value y_i for user agent $i \in \mathcal{N}$ is derived from its target critic network, which is updated as a delayed copy of the corresponding online critic network. The parameters $\boldsymbol{\omega}_i$ of the i-th online critic network are optimized via gradient descent to minimize the TD error. As illustrated in Fig. 5.2, the critic network $Q_i(\mathbf{S}^u, \mathbf{A}^u \mid \boldsymbol{\omega}_i)$ guides the policy improvement of the associated actor network $\pi_i(\mathbf{s}_i^u \mid \boldsymbol{\theta}_i)$. The actor parameters $\boldsymbol{\theta}_i$ are updated in the gradient ascent direction to maximize the expected Q-value. The actor $\pi_i(\mathbf{s}_i^u \mid \boldsymbol{\theta}_i)$ relies solely on the local observation $\mathbf{s}_i^u$, thereby eliminating the need for inter-user communications and enabling efficient decentralized execution.

5.4.3 Parallelized MADRL via Action Estimation

The primary challenge in applying MADDPG to low-level user agents lies in its requirement for an extensive dataset aggregated from all users, which renders it impractical for deployment in resource-constrained IoT networks. A second critical limitation stems from the centralized training paradigm, which relies on the joint

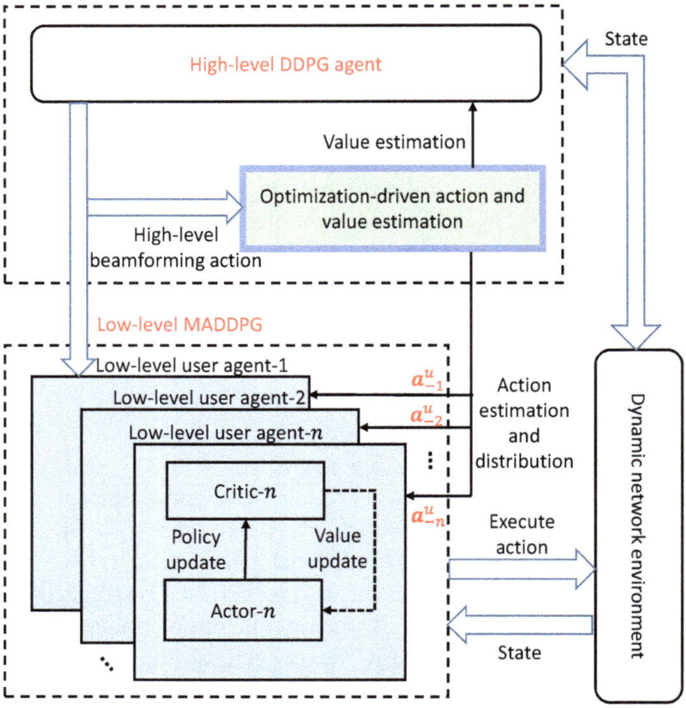

Fig. 5.2 The optimization-driven H-MADDPG framework, including an outer-loop DDPG for AP's high-level policy and an inner-loop optimization-driven MADDPG for user coordination. The AP selects beamforming strategies, computes model-based action estimates, and broadcasts them for users' policy initialization. Each user operates as an independent agent optimizing individual's offloading policy based on local state

state-action space of all agents. This leads to a rapid growth in input dimensionality, resulting in prohibitively slow training convergence and poor scalability. To address these issues, we enhance the learning performance of low-level MADDPG by synergistically combining the sample efficiency of model-based optimization methods with the robustness of the model-free DDPG framework. Specifically, conditioned on the AP's beamforming strategy, the high-level agent constructs a lightweight model to estimate the actions of the user agents, i.e., their offloading time and workload allocation strategies. This estimation is obtained by solving an approximate optimization problem under a fixed beamforming policy. The resulting action estimates, derived from global information available at the high-level agent, are then broadcast to all low-level user agents to guide their policy learning toward higher rewards. In the following subsections, we first present the optimization-based solution for time and workload allocation at the user level. Subsequently, we detail the architecture and training mechanism of the proposed optimization-driven MADDPG algorithm for low-level user agents.

Algorithm 1 Optimization-driven H-MADDPG for Hybrid MEC Offloading

Initialize DNNs and experience replay buffers of high-level and low-level agents
%**High-level AP's decision making**
Update beamforming action $\mathbf{a}^o$ by DDPG algorithm
Estimate low-level users' actions $\mathbf{a}^u$ and the target value y^* by solving (5.10)
Distribute $(\mathbf{a}^o, \mathbf{a}^u)$ to all low-level user agents
%**Low-level user's decision making**
for each low-level user agent $i \in \mathcal{N}$
% Fix other agents' action estimations $\mathbf{a}_{-i}^u$
    Update its own action $\tilde{\mathbf{a}}_i^u$ by DDPG algorithm
    Execute $\tilde{\mathbf{a}}_i^u$, observe reward, and buffer transitions
    Sample a mini-batch from the memory replay buffer
    Update the i-th critic and actor networks
end for
User agents report actions $\tilde{\mathbf{a}}^u = [\tilde{\mathbf{a}}_1^u, \tilde{\mathbf{a}}_2^u, \ldots, \tilde{\mathbf{a}}_N^u]$
AP estimates the target y of the joint action $(\mathbf{a}^o, \tilde{\mathbf{a}}^u)$
if y is greater than the optimization-driven target y^*
    AP updates the actions $(\mathbf{a}^o, \tilde{\mathbf{a}}^u)$ with prob. $1-\epsilon$ **else**
    AP updates the actions $(\mathbf{a}^o, \mathbf{a}^u)$ with prob. $1-\epsilon$
end if
Execute joint action, observe reward, and buffer transitions
Sample a mini-batch from the AP's replay buffer
Update the AP's critic and actor networks
Loop back to line 3

With the fixed beamforming strategy $\mathbf{w}_i$, the action estimation for each user corresponds to the solution of the energy minimization problem in (5.10). Both the objective and the workload completion constraints in (5.10) are linear with respect to the time and workload allocation variables $\{(\boldsymbol{t}_i, \boldsymbol{\ell}_i)\}_{i \in \mathcal{N}}$. Under this fixed beamforming policy, the harvested energy $e_{j,i} = \eta p_A |\mathbf{h}_j^H \mathbf{w}_i|^2$ becomes a known constant, rendering the energy budget constraint in (5.9) linear in the passive offloading time allocation $t_{b,i}$. Although the time and workload allocation pair $(t_{a,i}, \ell_{a,i})$ associated with active offloading exhibits nonlinear coupling in (5.9), it can be readily verified that the resulting MEC offloading problem admits a convex reformulation. Consequently, an efficient optimization module can be constructed to accurately estimate the offloading decisions of each user agent.

The action estimation provides an approximate initialization for the user agents' offloading decisions. However, due to dynamic channel conditions and environmental uncertainties, this estimation may not be optimal for individual users. To address this, each user is permitted to refine its action starting from the estimated one. Specifically, every user agent $i \in \mathcal{N}$ iteratively updates its local action $\mathbf{a}_i^u$ using the standard DDPG algorithm, under the assumption that all other agents adhere to the AP's action estimation. The overall workflow of the proposed optimization-driven MADDPG framework is depicted in Fig. 5.2. This design circumvents the need for frequent inter-agent communication and enables each user to perform policy updates independently based solely on its local observations. Once all user agents converge to their refined actions $\tilde{\mathbf{a}}_i^u$, they report these decisions back to the AP, which in turn leverages them to update its global action estimation. The key features and contributions of the optimization-driven MADDPG framework are summarized as follows:

- Model-based action estimation: The high-level agent (i.e., the AP) generates an informed initial action estimate for each user by solving a model-based optimization problem. This estimation leverages global system information, such as channel states and workload demands, collected from all users, thereby capturing inter-user resource couplings. Given the AP's ample computational resources, this optimization can be solved efficiently and reliably, offering a significantly better initialization than random exploration, particularly during the early stages of learning.
- Independent and parallel learning: The low-level MADDPG framework is decomposed into multiple parallel DDPG agents. After receiving the AP's action estimate $\mathbf{a}_i^u$, each user $i \in \mathcal{N}$ treats the actions of others $\mathbf{a}_{-i}^u$ as fixed and optimizes its own action $\tilde{\mathbf{a}}_i^u$ independently. This decoupling substantially reduces the effective dimensionality of both the state and action spaces, thereby enhancing learning efficiency and scalability.
- Optimization-driven learning: By solving the model-based problem in (5.10), the AP not only obtains an action estimate but also computes an associated objective value, denoted as the optimization-driven target y^*. This value serves as a performance baseline during the learning process. Additionally, the AP evaluates the actual performance y based on the refined actions $\{\tilde{\mathbf{a}}_i^u\}$ reported by the users. If $y > y^*$, the AP is more likely to accept the users' updated actions. Conversely, if $y < y^*$ due to exploratory noise in early learning stages, the AP relies on its model-based estimate. This adaptive mechanism ensures stable and efficient policy improvement.

The optimization-driven H-MADDPG algorithm is formally detailed in Algorithm 1. The overall framework comprises two nested loops: an outer-loop DDPG that governs the high-level policy of the AP, and an inner-loop optimization-driven MADDPG that coordinates the low-level user agents. In lines 3–5 of Algorithm 1, the AP selects its beamforming strategy $\mathbf{a}^o$ using its high-level DDPG policy and subsequently computes an estimate $\mathbf{a}^u$ of the edge users' time and workload allocation strategies by solving the model-based optimization problem. This action estimate is then broadcast to all user agents and serves as the initialization point for their individual policy exploration, as elaborated in lines 7–13. Following the refinement of their policies, the low-level users report their updated actions $\tilde{\mathbf{a}}^u = [\tilde{\mathbf{a}}_1^u, \tilde{\mathbf{a}}_2^u, \ldots, \tilde{\mathbf{a}}_N^u]$ back to the AP. In lines 14–15, the AP evaluates these actions to compute the empirical target value y. Finally, as shown in lines 16–22, the AP compares this observed performance y against the optimization-driven baseline y^* to determine whether to accept the users' updated actions. This comparison informs the AP's policy update and governs the transition to the next state.

5.5 Numerical Evaluation

In this section, we evaluate the performance gains achieved by the proposed optimization-driven H-MADDPG framework under varying system parameters, including edge users' workload demands, channel conditions, and energy availability. We consider a specific network topology wherein an MEC-enabled AP equipped with $M = 3$ antennas is positioned at the origin $(0, 0)$ of the coordinate system. Three edge users are located at $(5, 5)$, $(3, 3)$, and $(0, 4)$, respectively. Each user is allocated a unit time slot for offloading and computation. The AP's transmit power is set to $p_A = 100$ mW and the constant circuit power consumption at each edge user is $p_o = 1$ μW, consistent with the parameter configuration adopted in [30]. The MEC service pricing parameters are set to $\mu_1 = 0.2$ and $\mu_2 = 0.4$. Workload arrivals at each edge user follow a finite-state Markov chain with a mean arrival rate of $\lambda = 40$. Wireless channels are modeled as independent complex Gaussian random variables with zero mean and unit variance. Path loss is governed by the log-distance model $L_i = 20 + 20\log(d_i)$ (dB), where d_i denotes the distance between the AP and the i-th user. The neural network architectures for the actor in the proposed DRL framework consists of three fully connected layers with sizes `fc(STATE_DIM, 128)`, `fc(128, 64)`, and `fc(64, ACTION_DIM)`, all using sigmoid activation functions, where `fc(m,n)` refers to a fully connected layer of dimension $m \times n$. It is trained with a learning rate of 0.001, a batch size of 64, and a replay memory capacity of 2000. The critic network processes state and action inputs through separate streams. The state path uses `fc(STATE_DIM, 128)` and the action path uses `fc(ACTION_DIM, 128)`, which are then combined and passed through a `relu` activation and a final `fc(128, 1)` layer. The training uses a reward discount factor of 0.5, a smoothing parameter of 0.1 for target updates, and an ϵ-greedy exploration probability of 0.5.

5.5.1 *Convergence of Optimization-Driven H-MADDPG*

We begin by evaluating the learning performance of the proposed optimization-driven H-MADDPG algorithm in comparison with a model-free H-MADDPG baseline, which employs the model-free MADDPG algorithm for low-level user agents. As illustrated in Fig. 5.3a, the optimization-driven H-MADDPG achieves significantly faster convergence and attains a substantially higher cumulative reward than its model-free counterpart. The reward curve of the optimization-driven variant exhibits a sharp increase during the early learning phase and stabilizes earlier, whereas the model-free H-MADDPG demonstrates slower and more volatile learning progress. Despite fluctuations inherent to both approaches, the optimization-driven H-MADDPG yields approximately a 50% improvement in average reward. This performance gain can be attributed to two key mechanisms. First, the optimization-driven H-MADDPG mitigates the complexity of inter-agent interactions by decoupling the

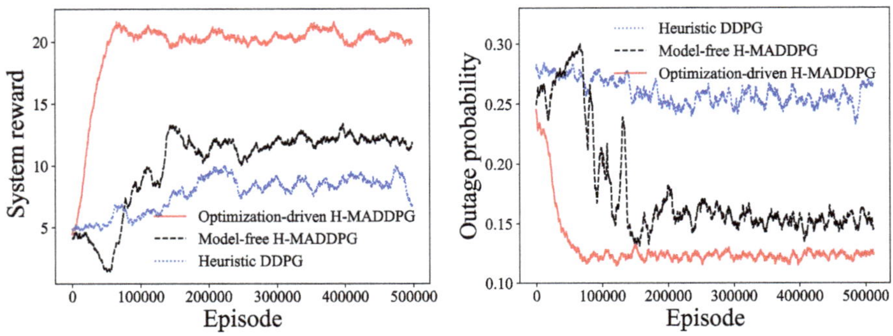

(a) Reward performance (b) Outage performance

Fig. 5.3 The optimization-driven H-MADDPG significantly outperforms model-free H-MADDPG, achieving 50% higher average reward with faster convergence and lower outage probability. Both H-MADDPG variants outperform heuristic DDPG in reward and reliability

low-level learning process. Given the AP's action estimation, each user agent operates as an independent DDPG learner, refining its offloading decision based solely on local observations. This parallelized and decentralized update scheme reduces the effective dimensionality of the joint action space and lowers the training overhead, thereby enhancing learning efficiency relative to the fully cooperative MADDPG framework. Second, the AP leverages its global view to compute not only an initial action estimate but also an optimization-driven target value y^*, which serves as a performance baseline or lower bound for each learning episode. This baseline guides the high-level policy toward actions that consistently yield higher rewards, effectively accelerating policy improvement. For further comparison, we also evaluate the optimization-driven H-MADDPG against a heuristic DDPG algorithm, as depicted in Fig. 5.3. In this heuristic approach, the AP updates its beamforming strategy via conventional DDPG, while anticipating user offloading decisions through a static optimization method. The results confirm that both H-MADDPG variants outperform the heuristic DDPG in terms of achieved reward, underscoring the benefits of integrating flexible learning framework with model-based optimization.

In Fig. 5.3b, we compare the outage probabilities of the two H-MADDPG variants. An outage event is defined as the failure to fully process a user's workload by the end of its allocated time slot, typically resulting from either excessive workload demands or insufficient energy resources. The results reveal that the optimization-driven H-MADDPG achieves a significantly lower outage probability compared to the model-free H-MADDPG. This improvement underscores the critical role of joint time allocation and workload partitioning between local computation and offloading in determining the system's overall processing capability. Consequently, the outage probability serves as a key indicator of the reliability of the users' offloading strategies. The reduced outage rate under the optimization-driven H-MADDPG demonstrates its enhanced robustness and adaptability to dynamic network conditions. Moreover, the sharp decline in outage probability during the early learning phase further corroborates the faster convergence behavior of the proposed algorithm. Although both reward and outage probability exhibit fluctuations during training, attributed to exploration noise and environmental stochasticity, the optimization-driven H-MADDPG stabilizes within a consistent performance range after approximately 10 K learning episodes. In contrast, the model-free variant converges more slowly and with greater variance. Additionally, both H-MADDPG algorithms achieve lower outage probabilities than the heuristic DDPG baseline, confirming that the hierarchical learning framework enables more effective and reliable offloading decisions.

In Fig. 5.4, we examine the reward dynamics of individual users within the optimization-driven H-MADDPG framework. Given the high-level AP's beamforming strategy, the low-level users concurrently refine their offloading decisions using the optimization-driven MADDPG algorithm, operating in parallel without direct coordination. Figure 5.4a illustrates the convergence behavior of the users' rewards in the low-level optimization-driven MADDPG. As the training progresses beyond 15 K episodes, the reward curves for all users exhibit a clear upward trend and eventually stabilize, confirming that the hierarchical learning architecture enables multiple agents to converge to consistent and effective policies. This result validates the feasibility and scalability of the proposed framework in multi-user settings. In Fig. 5.4(b), we zoom in on the reward trajectory of user-1 over several interaction rounds with the AP. Each round consists of two phases: (1) the AP updates its beamforming strategy based on its high-level policy, and (2) the low-level users adapt their offloading decisions using the MADDPG algorithm, initialized with the AP's action estimation. The AP leverages its global knowledge to compute an estimate of the optimal actions for all users and broadcasts this information, thereby enabling each user to operate as an approximately independent DDPG agent, as reflected in Fig. 5.4a. Once the low-level users reach stable offloading policies within a round, they report their refined actions back to the AP, which then uses this feedback to

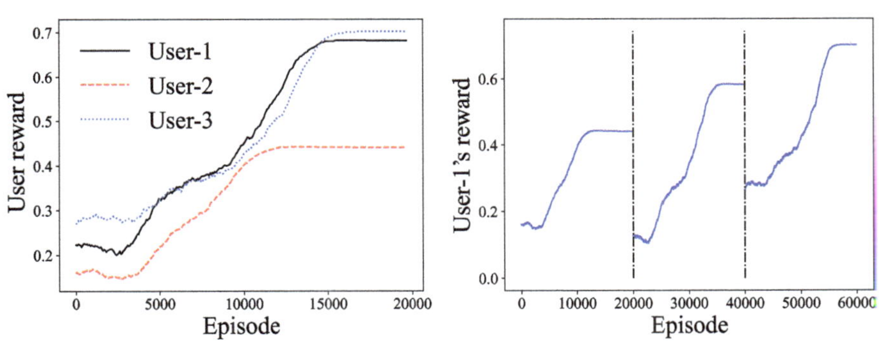

(a) User reward (b) System reward

Fig. 5.4 Reward dynamics of the optimization-driven H-MADDPG. All users' rewards increase and stabilize after 15 K episodes, validating the hierarchical framework's scalability. User-1's progressive reward improvement demonstrates effective policy evolution between high-level AP and low-level users

update its beamforming strategy in the subsequent rounds. As shown in Fig. 5.4b, user-1's reward improves progressively across rounds, demonstrating the complementary evolution of the high-level AP's policy and the low-level users' policies. This iterative refinement highlights the effectiveness of the hierarchical design in jointly optimizing system-wide performance through the integration of model-based optimization and model-free learning.

5.5.2 *Trade-Off Between Passive and Active Offloading*

In this section, we analyze the workload allocation among different offloading modes from the users' perspective. Passive offloading leverages the AP's signal beamforming, whereas active offloading consumes the users' energy budgets. The distribution of workload across local computation, active offloading, and passive offloading is primarily governed by the user's available energy budget. To further investigate this dependency, we examine how the edge user's energy budget influences its workload allocation strategy. When the AP's transmit power is low, the edge user operates under a constrained energy budget, leading it to suppress active offloading in favor of passive offloading, as illustrated in Fig. 5.5a. This strategy enhances opportunities for energy harvesting. Additionally, a significant portion of the workload is executed via local computation to satisfy the edge user's computational requirements. Conversely, as the AP's transmit power increases, the edge user can harvest more RF energy, thereby enabling a greater proportion of workload to be offloaded through active offloading.

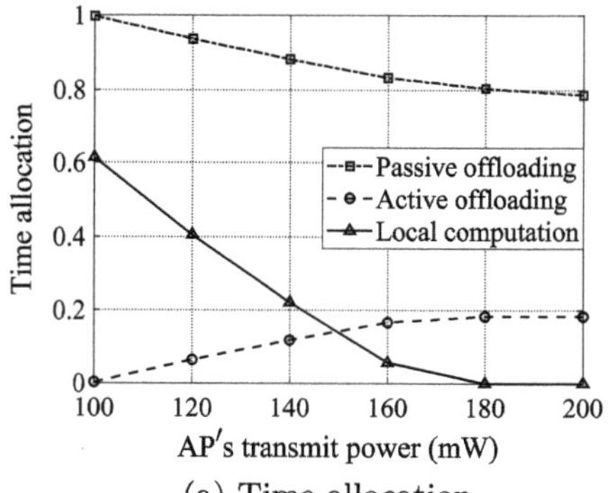

(a) Time allocation

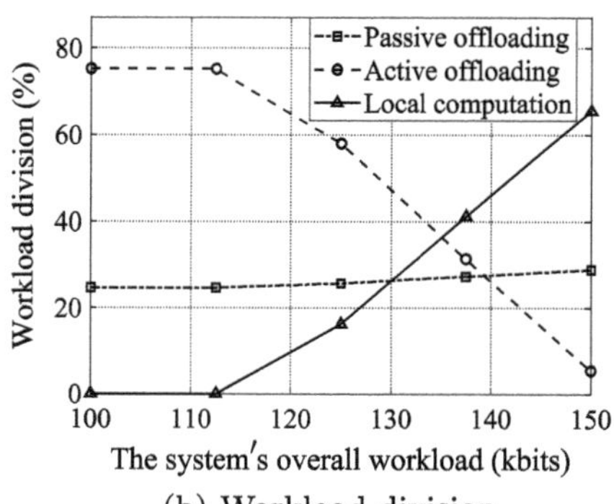

(b) Workload division

Fig. 5.5 Time and workload allocations with different computing schemes. Workload allocation depends on energy budget and AP's transmit power. Low power favors passive offloading and local computation due to energy constraints. Higher power enables more active offloading. Increased workload leads to more passive offloading and local computation as users balance energy harvesting and computational demands across offloading modes

In addition to the AP's transmit power, we also assess the impact of the system's total workload on workload allocation. As depicted in Fig. 5.5b, when the workload is relatively small, the edge user possesses sufficient energy to offload the entire workload to the MEC server via a combination of active and passive offloading. However, as the total workload increases, so do the energy demands of the edge users. Consequently, users allocate a larger share of their tasks to passive offloading to facilitate mutual energy supply, thereby reducing the time allocated to active offloading. To meet the growing computational demand under these constraints, local computation correspondingly increases, as evidenced in Fig. 5.5b.

5.5.3 *Hybrid Offloading Varies with Workload Demands*

In this section, we evaluate the performance of the hybrid offloading scheme under heterogeneous workload demands across individual users. Specifically, by increasing user-1's workload ℓ_1, we investigate how the proposed algorithm dynamically adjusts offloading decisions for different users to satisfy their respective energy and computational requirements. As ℓ_1 grows from 30 to 45 kbits, Fig. 5.6a illustrates the adaptive time allocation of user-2 among local computation, active offloading, and passive offloading. As anticipated, user-1 shifts an increasing portion of its workload to active offloading to accommodate the rising demand. Consequently, user-2 reduces its time allocation for active offloading, as shown in Fig. 5.6a, due to diminished energy harvesting opportunities during user-1's passive offloading slot $t_{b,1}$. As such, user-2 reallocates more workload to passive offloading. Additionally, user-2 increases its local computation load to mitigate the risk of workload outage. User-2's strategic adjustment simultaneously enhances the energy harvested by user-1, thereby supporting user-1's computational needs.

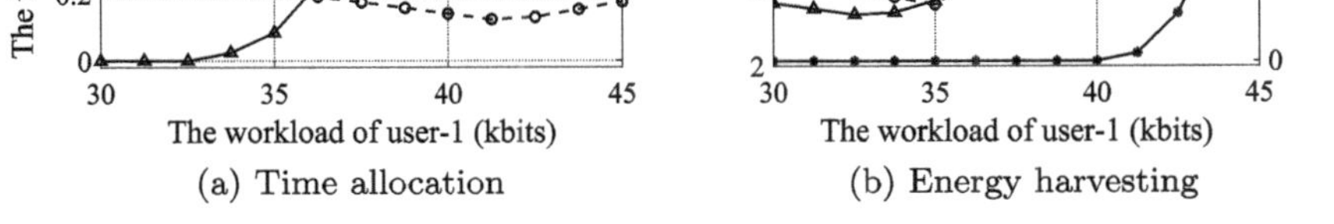

(a) Time allocation (b) Energy harvesting

Fig. 5.6 Offloading decisions vary with workload demands. As user-1's workload increases, user-2 adaptively re-allocates time among different offloading modes, reducing active offloading, increasing passive offloading and local computation. Outage occurs when user-1's workload exceeds 40 kbits

Figure 5.6b presents the energy harvesting capabilities of the users as ℓ_1 increases. A notable disparity is observed between user-1 and the other users (i.e., user-2 and user-3), highlighting the hybrid offloading scheme's ability to leverage complementary cooperation among users with heterogeneous roles. When ℓ_1 exceeds 40 kbits, workload outage becomes unavoidable, and the outage probability rises sharply, as depicted in Fig. 5.6b. Under such conditions, the MEC system fails to process user-1's entire workload within a single data frame, despite cooperative support from user-2 and user-3. Importantly, simulation results indicate that the outage probabilities of user-2 and user-3 remain unchanged as user-1's outage probability escalates. This suggests that the MEC system prioritizes resource reservation for user-2 and user-3, effectively safeguarding their service reliability at the expense of overall energy efficiency, which degrades as user-1's demand intensifies.

5.6 Discussions and Conclusions

We have studied a hybrid computation offloading scheme for multi-user MEC systems, leveraging a hierarchical MADRL framework to maximize energy efficiency. The hierarchical learning architecture decouples control into a high-level agent at the AP, responsible for beamforming design, and multiple low-level agents at the edge users, which subsequently optimize individuals' offloading and resource allocation decisions using a MADDPG framework. To enhance learning efficiency, we further design an optimization-driven learning algorithm that integrates analytical insights into the low-level policy updates, incorporating domain knowledge to improve sample efficiency. Numerical results demonstrate that the proposed framework significantly improves convergence, reliability, and reward performance compared to model-free methods. These findings highlight the effectiveness of hierarchical learning in balancing global coordination with local adaptability, offering a scalable and intelligent control for dynamic and resource-constrained MEC systems.

References

1. H. Zhou, Y. Long, S. Gong, K. Zhu, D.T. Hoang, D. Niyato, Hierarchical multi-agent deep reinforcement learning for energy-efficient hybrid computation offloading. IEEE Trans. Veh. Technol. **72**(1), 986–1001 (2023)
2. H. Zhou, Y. Long, W. Zhang, J. Xu, S. Gong, Hierarchical multi-agent deep reinforcement learning for backscatter-aided data offloading, in *Proceedings of IEEE WCNC* (2022), pp. 542–547
3. A. Al-Fuqaha, M. Guizani, M. Mohammadi, M. Aledhari, M. Ayyash, Internet of things: a survey on enabling technologies, protocols, and applications. IEEE Commun. Surv. Tut. **17**(4), 2347–2376 (2015)
4. H. Guo, J. Li, J. Liu, N. Tian, N. Kato, A survey on space-air-ground-sea integrated network security in 6G. IEEE Commun. Surv. Tut. **24**(1), 53–87 (2022)

5. Y. Mao, C. You, J. Zhang, K. Huang, K.B. Letaief, A survey on mobile edge computing: the communication perspective. IEEE Commun. Surv. Tut. **19**(4), 2322–2358 (2017)
6. Z. Ning, Y. Yang, X. Wang, L. Guo, X. Gao, S. Guo, G. Wang, Dynamic computation offloading and server deployment for UAV-enabled multi-access edge computing. IEEE Trans. Mobile Comput. **22**(5), 2628–2644 (2023)
7. X. Lu, D. Niyato, H. Jiang, D.I. Kim, Y. Xiao, Z. Han, Ambient backscatter assisted wireless powered communications. IEEE Wireless Commun. **25**(2), 170–177 (2018)
8. G. Yang, D. Yuan, Y.-C. Liang, R. Zhang, V.C.M. Leung, Optimal resource allocation in full-duplex ambient backscatter communication networks for wireless-powered IoT. IEEE Internet Things J. **6**(2), 2612–2625 (2019)
9. P.X. Nguyen, D.-H. Tran, O. Onireti, P.T. Tin, S.Q. Nguyen, S. Chatzinotas, H. Vincent Poor, Backscatter-assisted data offloading in OFDMA-based wireless-powered mobile edge computing for IoT networks. IEEE Internet Things J. **8**(11), 9233–9243 (2021)
10. F. Wang, X. Zhang, Joint optimization for traffic-offloading and resource-allocation over RF-powered backscatter wireless networks. IEEE J. Sel. Topic. Signal Process. **15**(5), 1127–1142 (2021)
11. Y. Zhang, B. Di, Z. Zheng, J. Lin, L. Song, Distributed multi-cloud multi-access edge computing by multi-agent reinforcement learning. IEEE Trans. Wireless Commun. **20**(4), 2565–2578 (2021)
12. N.C. Luong, D.T. Hoang, S. Gong, D. Niyato, P. Wang, Y.C. Liang, D.I. Kim, Applications of deep reinforcement learning in communications and networking: a survey. IEEE Commun. Surv. Tut. **21**(4), 3133–3174 (2019)
13. S. Gong, Y. Xie, J. Xu, D. Niyato, Y.-C. Liang, Deep reinforcement learning for backscatter-aided data offloading in mobile edge computing. IEEE Netw. **34**(5), 106–113 (2020)
14. L. Ale, N. Zhang, X. Fang, X. Chen, S. Wu, L. Li, Delay-aware and energy-efficient computation offloading in mobile edge computing using deep reinforcement learning. IEEE Trans. Cogn. Commun. Netw. **7**(3), 881–892 (2021)
15. M. Min, L. Xiao, Y. Chen, P. Cheng, D. Wu, W. Zhuang, Learning-based computation offloading for IoT devices with energy harvesting. IEEE Trans. Veh. Technol. **68**(2), 1930–1941 (2019)
16. X. Chen, H. Zhang, C. Wu, S. Mao, Y. Ji, M. Bennis, Optimized computation offloading performance in virtual edge computing systems via deep reinforcement learning. IEEE Internet Things J. **6**(3), 4005–4018 (2019)
17. J. Liu, H. Guo, J. Xiong, N. Kato, J. Zhang, Y. Zhang, Smart and resilient EV charging in sdn-enhanced vehicular edge computing networks. IEEE J. Sel. Areas Commun. **38**(1), 217–228 (2020)
18. Y. Wei, F.R. Yu, M. Song, Z. Han, Joint optimization of caching, computing, and radio resources for fog-enabled IoT using natural actor-critic deep reinforcement learning. IEEE Internet Things J. **6**(2), 2061–2073 (2019)
19. J.N. Foerster, Y.M. Assael, N. De Freitas, S. Whiteson, Learning to communicate with deep multi-agent reinforcement learning, in *Proceedings of International Conference on Neural Information Processing Systems (NeurIPS)*, Barcelona, Spain (2016), pp. 2145–2153
20. V. Mnih, A. P. Badia, L. Mirza, A. Graves, T. Harley, T. P. Lillicrap, D. Silver, K. Kavukcuoglu, Asynchronous methods for deep reinforcement learning, in *Proceedings of International Conference on Machine Learning (ICML)*, New York, USA (2016), pp. 2850–2869
21. R. Lowe, Y. Wu, A. Tamar, J. Harb, P. Abbeel, I. Mordatch, Multi-agent actor-critic for mixed cooperative-competitive environments, in *Proceedings of International Conference on Neural Information Processing Systems (NeurIPS)*, Long Beach, USA (2017), pp. 6382–6393
22. X. Huang, S. Leng, S. Maharjan, Y. Zhang, Multi-agent deep reinforcement learning for computation offloading and interference coordination in small cell networks. IEEE Trans. Veh. Technol. **70**(9), 9282–9293 (2021)
23. L. Wang, K. Wang, C. Pan, W. Xu, N. Aslam, L. Hanzo, Multi-agent deep reinforcement learning-based trajectory planning for multi-UAV assisted mobile edge computing. IEEE Trans. Cogn. Commun. Netw. **7**(1), 73–84 (2021)

24. C. Liu, F. Tang, Y. Hu, K. Li, Z. Tang, K. Li, Distributed task migration optimization in MEC by extending multi-agent deep reinforcement learning approach. IEEE Trans. Parallel Distrib. Syst. **32**(7), 1603–1614 (2021)
25. Z. Ning, P. Dong, X. Wang, M.S. Obaidat, X. Hu, L. Guo, Y. Guo, J. Huang, B. Hu, Y. Li, When deep reinforcement learning meets 5G-enabled vehicular networks: A distributed offloading framework for traffic big data. IEEE Trans. Ind. Inform. **16**(2), 1352–1361 (2020)
26. N. Zhu, X. Xu, S. Han, S. Lv, *Sleep-Scheduling and Joint Computation-Communication Resource Allocation in MEC Networks for 5G IIoT* in Proceedings of IEEE WCNC, Nanjing, China (2021), pp. 1–7
27. X. Wang, Z. Ning, S. Guo, M. Wen, L. Guo, H.V. Poor, Dynamic UAV deployment for differentiated services: a multi-agent imitation learning based approach. IEEE Trans. Mobile Comput. **22**(4), 2131–2146 (2023)
28. H. Guo, J. Liu, UAV-enhanced intelligent offloading for internet of things at the edge. IEEE Trans. Ind. Inform. **16**(4), 2737–2746 (2020)
29. L. Qian, Y. Wu, F. Jiang, N. Yu, W. Lu, B. Lin, NOMA assisted multi-task multi-access mobile edge computing via deep reinforcement learning for industrial internet of things. IEEE Trans. Ind. Inform. **17**(8), 5688–5698 (2021)
30. S. Gong, Y. Zou, J. Xu, D.T. Hoang, B. Lyu, D. Niyato, Optimization-driven hierarchical learning framework for wireless powered backscatter-aided relay communications. IEEE Trans. Wireless Commun. **21**(2), 1378–1391 (2022)
31. F. Zhou, R.Q. Hu, Computation efficiency maximization in wireless-powered mobile edge computing networks. IEEE Trans. Wireless Commun. **19**(5), 3170–3184 (2020)
32. S. Guo, B. Xiao, Y. Yang, Y. Yang, *Energy-Efficient Dynamic Offloading and Resource Scheduling in Mobile Cloud Computing* in Proceedings of IEEE INFOCOM, San Francisco, USA (2016), pp. 1–9
33. S. Gong, L. Gao, J. Xu, Y. Guo, D.T. Hoang, D. Niyato, Exploiting backscatter-aided relay communications with hybrid access model in device-to-device networks. IEEE Trans. Cogn. Commun. Netw. **5**(4), 835–848 (2019)
34. H. Zhou, K. Jiang, X. Liu, X. Li, V.C.M. Leung, Deep reinforcement learning for energy-efficient computation offloading in mobile edge computing. IEEE Internet Things J. **9**(2), 1517–1530 (2022)
35. M. Li, J. Gao, L. Zhao, X. Shen, Adaptive computing scheduling for edge-assisted autonomous driving. IEEE Trans. Veh. Technol. **70**(6), 5318–5331 (2021)

Chapter 6
Hierarchical MADRL for UAV-Assisted Wireless Networks

6.1 Case Study in UAV-Assisted Wireless Networks

The integration of UAVs into wireless networks offers a promising solution for connecting massive number of wireless devices on the ground over a large geographic area, constituting a critical component of future IoT systems, e.g., [1, 2]. This is especially critical for low-cost IoT devices with hardware constraints, making it difficult to meet the QoS requirements for reliable data transmissions, due to energy shortage, limited channel capacity, and NLoS channel conditions [3]. These practical challenges can be alleviated by deploying UAVs as the aerial and mobile access points to assist data transmissions from the GUs to remote BSs or edge servers [4]. By leveraging UAVs' capabilities for fast deployment, high mobility, and flexibility, they can be integrated in UAV-assisted wireless networks to serve GUs distributed over a large area beyond the direct coverage of a traditional BS. This not only extends the communication range but also improves the overall network capacity for dense IoT devices. However, the realization of such performance gain in UAV-assisted wireless networks requires the optimization of the UAVs' networking and mobility control strategies jointly with the GUs' transmission control, including UAV-GU association, access scheduling, collaborative beamforming, and power control. The tight spatio-temporal couplings among control variables create complex and high-dimensional network optimization problems, which are hardly tractable by conventional optimization methods.

In this case study, we examine the customized design and applications of MADRL for UAV-assisted wireless networks, consisting of multiple UAVs serving a large number of GUs. Specially, multiple UAVs are coordinated by the remote BS and deployed to forward GUs' sensing information to the remote BS. Individual UAVs can dynamically plan their trajectories and network formation strategy to improve efficiency for their collaborative data transmissions. The UAVs' trajectory planning aims at collecting all GUs' data, while the network formation adapts the UAVs' multi-hop relaying network to minimize the energy consumption and transmission delay.

S. Gong et al., *Optimization-Driven Deep Reinforcement Learning for Wireless Networks*, Wireless Networks, https://doi.org/10.1007/978-3-032-22997-7_6

Such a collaborative strategy can improve data rates and reduce energy consumption of distant GUs far from the BS. It is clear that network formation strategy should be dynamically updated according to the UAVs' trajectories, which determines the channel conditions between every pair of UAVs. This implies that the joint network formation and trajectory planning problem is complicated with spatio-temporal couplings among UAVs, and we propose a two-step iterative approach to solve it. Firstly, we devise a heuristic algorithm to dynamically update the UAVs' network formation, which aims at balancing the UAVs' energy consumption and data queue size. Then, maintaining the same network formation, we further search the UAVs' trajectories by using model-free MADRL to interact with the network environment. The model-free MADRL normally relies on centralized training and distributed execution scheme, which may require excessive training samples and have slow learning performance. In this case study, we improve MADRL's learning efficiency by employing Bayesian optimization to estimate the UAVs' trajectory points based on historical information. The bayesian estimation aims to improve the size of data collected from the GUs. Instead of random action exploration, the bayesian estimated action can help avoid inefficient action explorations and thus improve the learning performance. The detailed analysis and algorithm design in the following sections were originally presented in our earlier work [5]. This chapter refines and reorganizes that content to ensure consistency, clarity, and coherence within the scope of this book.

6.1.1 Motivations and Contributions

Leveraging the superior G2U direct channel conditions, UAVs can firstly hover at some points of interest, collect sensing information from GUs, and subsequently help forward the sensing information to the remote BS. By strategically planning the UAVs' trajectories and transmission control strategies, we can enhance the data rate and reduce the transmission latency for all GUs. This can improve the overall network throughput and service capacity [6]. GUs can achieve a higher data rate and lower transmit power by uploading sensing information to nearby UAVs instead of direct communications with the remote BS. In addition to capacity and efficiency improvement, UAVs can also plan their trajectories to enhance freshness of GUs' sensing information and thus reduce the average AoI to collect all GUs' sensing data timely [7].

However, it becomes a high-dimensional control problem to optimize UAVs' trajectories in a multi-UAV-assisted wireless network, complicated by intricate spatio-temporal couplings among different UAVs and GUs. The complexity is further compounded by UAVs' limited energy budgets, physical flight restrictions, and the GUs' diverse service demands. For instance, while deploying multiple UAVs can expands the service coverage, this increases energy consumptions for distant UAVs to report or carry their sensing information back to the remove BS. Alternatively, it is also possible to employ multiple UAVs to serve simultaneously the same set of GUs, which can increase the GUs' channel access probabilities and the sensing capacity.

However, this may sacrifice the coverage range and transmission capacity. Due to close proximity, the UAVs' mutual interference also becomes severe and demands more sophisticated transmission scheduling. The above analysis implies that each UAV's trajectory is interdependent, affecting not only its own energy and information efficiency but also the other UAVs' current and future trajectories. These complex spatio-temporal couplings make the joint optimization of UAVs' trajectories and transmission control strategies particularly challenging for serving a large group of GUs. The UAVs' cooperation can be enabled by allowing the U2U direct communications when two UAVs are in close proximity with preferable channel conditions, e.g., [8, 9], which facilitates the formation of a multi-hop relaying network of UAVs. The UAVs' network formulation allows each UAV to firstly collect and cache GUs' data, and then opportunistically forward it to a neighboring UAV when their U2U channel becomes preferable for data transmissions. Moreover, with the change of UAVs' trajectories, the UAVs' network formation strategy should also be made adpative allowing UAVs to adaptively switch U2U links based on real-time channel conditions, energy status, and data queue sizes.

Currently, the UAVs' joint trajectory optimization and network formation have not been extensively studied for UAV-assisted wireless networks. Given a predetermined network formation strategy, the UAVs' trajectory optimization can be discretized in sequential time slots and formulated as a mobility control problem to find the optimal trajectory point in each time slot [10]. However, conventional optimization methods often require complete global information, such as GUs' distribution and traffic demands, which is impractical in dynamic network environment. Model-free DRL method presents a promising and alternative solution to adapt the UAVs' trajectories by direct interaction with the wireless environment, e.g., [11, 12]. A typical DRL implementation often employs a centralized controller at the BS to collect global state information and compute optimal actions for all UAVs. However, this may introduce substantial communication and computational overhead, especially as the number of UAVs grows in large-scale wireless networks. Alternatively, a decentralized multi-agent implementation can view each UAV as an independent decision maker, which can adapt its own trajectory based on local observations of the network environment. Although the multi-agent implementation reduces operational overhead, still it relies on a resource-intensive centralized training process, which can be costly especially when the UAVs' information exchange are not always available. These challenges necessitate the development of more efficient algorithmic frameworks capable of jointly optimizing the UAVs' trajectories and network formation to fulfill the GUs' traffic demands reliably and dynamically.

In this case study, focusing on a multi-UAV-assisted wireless network, we aim at minimizing the overall time delay and energy efficiency for GUs' data transmissions by jointly optimizing GUs' transmission control, UAVs' trajectory planning and network formation strategies. Our analysis reveals that the UAVs' trajectories and network formation are closely dependent on each other, though they are previously studied as two separate problems in the literature [8]. To serve a large-scale wireless network, UAVs' trajectories are highly dependent on the GUs' spatial distribution and traffic demands. As UAVs adapt their trajectories to serve these dynamic demands, they may fly far apart and result in deteriorated or disconnected U2U communication links. This necessitates the design of an adaptive network formation that can dynamically evolve with the UAVs' changing trajectories. To address this coupled problem, we propose a novel two-step iterative algorithm that jointly optimizes UAVs' trajectories and network formation. The core idea of our adaptive network formation strategy is to continuously evaluate the UAVs' resource consumption (e.g., energy status, data backlog), which are used to assess the quality of the current network formation strategy and then drive its adaptation. Specifically, if the current trajectories in a given time slot lead to unstable connections or significantly unbalanced resource consumption, the BS triggers the adaptation of network formation that tries to ensure load balance among different UAVs. The adaptive network formation will optimize the UAVs' network topology (e.g., U2U connections) or channel allocation to rectify the resource imbalance. Such an adapted network formation will in turn provide a new connectivity constraint for UAVs' trajectory planning. We then employ the MADDPG algorithm to enable each UAV to update its trajectory given the new network formation strategy [13]. These two processes iterate to maintain a balanced and efficient wireless network. Furthermore, to enhance learning efficiency, we introduce an action estimation mechanism in the MADDPG framework by using Bayesian optimization to estimate more rewarding actions for each UAV. The optimization-drive action estimation can prune ineffective action exploration and accelerating convergence of the MADDPG framework. Extensive simulations demonstrate that the proposed Bayesian-enhanced MADDPG framework for iterative trajectory planning and adaptive network formation significantly enhances overall network performance, exhibiting superior learning efficiency, convergence stability, and final reward performance over the conventional MADDPG approaches.

In summary, the main contributions of this case study are as follows:

- Multi-UAV-assisted wireless network: we employ a set of collaborative UAVs to collect data from GUs and forward it to a remote BS by U2U multi-hop relay communications. To balance the UAVs' traffic load and resource consumption, the UAVs' network formation strategy should be adaptively updated according to the UAVs' trajectories. This motivates us to propose a novel framework for the joint optimization of the UAVs' trajectories and network formation to exploit the UAVs' cooperative gain.
- Trajectory-triggered network formation: We propose a two-step iterative algorithm that alternately optimizes the UAVs' trajectories and their network formation. When the current trajectories result in unbalanced traffic load and resource consumption, according to the UAVs' residual energy, channel conditions, and data queue sizes, the adaptation of network formation is then triggered to optimize the U2U connections, aiming to redistribute the traffic load and restore balance across different UAVs.
- Bayesian-enhanced MADDPG for trajectory planning: Given the network formation, multiple UAVs' trajectory planning is formulated into the MADDPG framework, which is integrated with a Bayesian optimization module to improve its learning efficiency. The Bayesian module estimates promising flying locations

based on historical trajectory points, thereby guiding the DRL's action exploration toward more rewarding policies. The DRL's action exploitation in return enriches the Bayesian inference model, enabling increasingly accurate action estimations over time. Extensive simulation results confirm that this integrated learning framework achieves significant improvements in learning speed, stability, and reward performance compared to the conventional MADDPG algorithms.

6.1.2 Related Works

6.1.2.1 Multi-UAV Cooperative Networks

Considering the UAVs' fast deployment, mobility, and flexibility, the optimization of the UAVs' trajectories becomes one of the main technical problems in UAV-assisted wireless networks to achieve diverse network control objectives, such as service coverage expansion, network throughput maximization, and transmission delay minimization. Most of the existing research works considered a centralized network structure in which the UAVs only have direct communications with the remote BSs and aimed to optimize individual UAVs' trajectories. For instance, the authors in [14] optimized the UAV's placement location to maximize the number of GUs under its service coverage. Without U2U relay communications, the authors in [15] studied the coalition formation to optimize the UAVs' task assignment in a large service area. Each UAV is assigned with a different service area to avoid resource conflict among each other. However, for large-scale wireless networks, direct communications between UAVs and BS may become unreliable and thus limit the practical service range of multi-UAV-assisted wireless networks.

UAVs' collaboration can be realized by enabling multi-hop U2U relay communications to expand the network coverage and increase the sensing capability for distant GUs [16]. The authors in [17] formulated a network formation game to construct the UAV backhaul that helps forward GUs' sensing information to the BS by multi-hop relay communications among different UAVs. The authors in [8] further improved the throughput performance of the network formation by a joint optimization of the UAVs' channel assignment and flight control. A multi-hop UAV-assisted relay network was proposed in [18] to assist a set of terrestrial transceivers to communicate with each other, aiming to maximize the minimum transmission rate by jointly optimizing UAVs' deployment locations, transmit power, and bandwidth allocation. A similar multi-hop UAV-assisted task offloading system was studied in [9], aiming to maximize the on-the-fly computation capability by jointly optimizing the UAVs' resource allocation and deployment strategies. To address the multi-UAV routing problem in MEC systems, the authors of [19] jointly optimized computation offloading and multi-hop routing strategies. Aiming to save UAVs' energy consumption, the routing strategy proposed in [20] allows each UAV to either rely on hop-by-hop routing algorithm or a cooperative coalition formation game. The choice between them is adaptive, depending on the UAVs' data demands, locations, and network topology. These approaches necessitate adaptive network formation based on real-time network status, such as network topology, channel conditions, and energy levels, enabling UAV-assisted wireless networks to manage complex sensing tasks and meet diverse requirements.

6.1.2.2 DRL for UAV-Assisted Wireless Networks

DRL enables the UAV agents to dynamically adjust their trajectories and data offloading strategies by interacting with the environment, corresponding to time-varying workload demands and channel conditions. This approach has been applied in various scenarios. For instance, the authors in [21] employed an actor-critic DRL method to optimize data collection using multiple UAVs. Their strategy first aggregates GUs via K-means clustering and then optimizes the UAVs' trajectories through the actor-critic DRL algorithm. The authors in [22] proposed DDPG to jointly adapt UAVs' trajectories and transmission scheduling. Beyond throughput enhancement, transmission delay and AoI can be significantly reduced by the rapid deployment of UAV-assisted wireless networks. The focus on AoI is explored in [23], which studied its minimization in a UAV-assisted network with RF-based wireless power transfer. The authors used a DQN approach to adapt the UAV's trajectory alongside GUs' scheduling and energy harvesting policies. Further, the authors in [24] addressed the trade-off between AoI and energy consumption, minimizing a weighted sum of average AoI, UAV's propulsion energy, and GUs' transmission energy by using the TD3 algorithm to adapt the UAV's speed, hovering locations, and bandwidth allocation.

The model-free MADRL framework has been effectively applied in literature to adapt UAVs' trajectories for data collection in complex wireless network environment. A common approach is to view each UAV as an independent agent making decisions based on local observations. This is seen in works utilizing multi-agent DQN for coverage and capacity maximization [25, 26]. The MADDPG algorithm has been used for joint target assignment and trajectory planning in [27] and to ensure secure communication in UAV-assisted wireless networks [28]. Furthermore, the authors in [29] employed a multi-agent actor-critic method to control UAVs' mobility, compute resource allocation, and offloading scheduling. For an energy-constrained system in [30], each UAV can plan its route to collect GUs' sensing data and fly back to a charging station on-demand when its energy supply becomes insufficient. The authors in [30] introduced a novel multi-agent deep option learning (MADOL) algorithm to minimize UAVs' total flight time by adapting the UAV-GU association and charging scheduling strategies. Several studies focused on AoI minimization in UAV-assisted wireless networks to ensure freshness of information for randomly generate status updates from GUs. A actor-critic DRL approach was used in [31] to adapt each UAV's sensing and flying decision, whereas MADDPG was proposed in [32] to jointly optimize UAVs' trajectories and transmission control strategies for AoI reduction.

6.2 System Model and Problem Formulation

We consider a multi-UAV-assisted wireless network consisting of one BS and a set of UAVs denoted by $\mathcal{N} = \{1, 2, \ldots, N\}$. There is also a set of ground-based sensors or IoT user devices, denoted by $\mathcal{M} = \{1, 2, \ldots, M\}$, which are spatially distributed and may lie outside the direct communication range of the BS. The UAVs are deployed to fly within the coverage area and collect sensing data from these GUs. The collected data is temporarily buffered at the UAVs before being forwarded to the BS, particularly when the channel conditions become favorable. Each UAV has a limited buffer capacity $D_{\max}$, and a buffer outage occurs if the accumulated data exceeds this limit. Depending on their channel conditions and energy status, each UAV can either communicate directly with the BS or relay its data through other UAVs. The direct inter-UAV communications establish a multi-hop network topology, referred to as the network formation, that enables cooperative data forwarding. This multi-hop relaying structure has the potential to reduce both the overall transmission delay and energy consumption.

6.2.1 *Time-Slotted Fly-Sense-Offload Protocol*

We adopt a time-slotted structure for the UAVs' operations. The set of time slots is denoted by $t \in \mathcal{T} \triangleq \{1, 2, \ldots\}$. In each time slot $t \in \mathcal{T}$, a UAV can adjust its position, collect sensing data from GUs, and offload buffered data to either the BS or another UAV. Each time slot has unit duration and is divided into three sequential sub-slots,i.e., the flying sub-slot t_f, the sensing sub-slot t_s, and the offloading sub-slot t_o. During the flying sub-slot t_f, each UAV moves from its current location to a new position at a controllable speed v_i. In the sensing sub-slot t_s, the UAV collects data from the GU with the strongest sensing signal within its service coverage. Each GU $m \in \mathcal{M}$ has a fixed data transmission demand W_m. A GU stops transmitting once its remaining data is fully delivered or when the UAV's buffer becomes full. In the offloading sub-slot t_o, each UAV can transmit a portion of its buffered data either directly to the BS or to a nearby UAV with better channel conditions and sufficient energy. During this phase, the UAV also updates its status information to the BS, including its current location, buffer occupancy, energy level, and channel state. This information enables the BS to optimize UAVs' trajectories and transmission control strategies. For simplicity, we assume a fixed sensing duration t_s, while the duration of the flying sub-slot t_f and the offloading sub-slot t_o are jointly optimized to maximize transmission efficiency. The proposed time-slotted Fly-Sense-Offload protocol enables multi-UAV-assisted data collection and supports three types of communication links among GUs, UAVs, and the BS, as detailed below.

- The G2U channels are used for UAVs to collect sensing data from GUs. Due to physical obstacles, direct channels from the GUs to the BS are assumed unavailable. All data must be relayed through UAVs. A GU can successfully transmit its data to a UAV only if the received SINR exceeds a predefined threshold.

- The U2B channels are used when a UAV communicates directly with the BS, typically when it is in close proximity. U2B communications occur over a set of dedicated reporting channels shared among all UAVs. The achievable data rate depends on the locations, transmit power, and channel conditions of UAVs.
- The U2U channels enable nearby UAVs to communicate and form a multi-hop relay network. When a UAV is far from the BS, it can forward data via intermediate UAVs to the BS by cooperative relaying.

We assume that each UAV is equipped with a single antenna. It can use either the U2B or U2U channel for transmission in a given time slot. The U2U and U2B channels share the same spectrum resources. To manage interference and limited bandwidth, all UAVs share a total of K orthogonal sub-channels, denoted by the set $\mathcal{K} = \{1, 2, \ldots, K\}$. Once a G2U link is established, the GU will upload its sensing data to the UAV during the sensing sub-slot t_s. After data collection, the UAVs will update their flight and communication strategies during the offloading sub-slot t_o.

6.2.2 UAVs' Network Formation

The U2U connections enable the formation of a multi-hop backhaul network, through which sensing data can be relayed from distant UAVs to the BS. We refer to this dynamic connectivity structure as the UAVs' network formation. Given a particular network formation, each UAV can optimize its flight trajectory and choose to forward buffered data either directly to the BS via U2B links or through neighboring UAVs via U2U relaying. However, as UAVs move to different hovering positions, the U2U channel conditions change, potentially rendering the current network formation suboptimal or obsolete. Therefore, the network formation must be jointly optimized with the UAVs' trajectories. It should be dynamically adapted based on real-time factors such as channel quality, available energy, buffer occupancy, and UAVs' locations. For instance, when UAVs are far from the BS, direct U2B transmissions may incur high delay and excessive energy consumption. In such cases, reconfiguring the network formation to exploit multi-hop U2U relaying communications can significantly improve efficiency. Moreover, when GUs generate uneven data traffic, i.e., some UAVs collect large amounts of data while others remain lightly loaded, the resulting traffic imbalance can lead to transmission congestion and increased delay. To mitigate this, UAVs with heavy buffer loads can offload data to neighboring UAVs using high-speed U2U links. In the following, we present the formal problem formulation and the resource constraints for network formation.

For notational convenience, we denote the BS as UAV-0, which is fixed at a known location. Let $\tilde{\mathcal{N}} = \mathcal{N} \cup \{0\}$ represent the extended set of all aerial nodes, including the N mobile UAVs and the BS (labeled as UAV-0). We define a binary matrix $\Phi(t) = [\phi_{i,j}^k(t)]_{i,j\in\tilde{\mathcal{N}},k\in\mathcal{K}}$ to represent the UAVs' network formation strategy at time slot t. Here, $\phi_{i,j}^k(t) = 1$ indicates that the k-th sub-channel is allocated to the U2U communication link from UAV-i to UAV-j. By definition, the UAV-0 does not participate in U2U relaying, so $\phi_{0,j}^k(t) = 0$ for all $j \in \tilde{\mathcal{N}}$. Additionally, self-connections are not allowed, hence $\phi_{j,j}^k(t) = 0$ for all $j \in \tilde{\mathcal{N}}$. Each sub-channel can be used for either data transmission or reception by a UAV but not both simultaneously due to half-duplex operation. Therefore, for any UAV-i and sub-channel k, the total number of incoming and outgoing U2U links must not exceed one. This requirement is captured by the following constraint:

$$\sum_{m \in \mathcal{N}, m \neq i} \phi_{m,i}^k(t) + \sum_{j \in \tilde{\mathcal{N}}, j \neq i} \phi_{i,j}^k(t) \leq 1, \quad \forall i \in \tilde{\mathcal{N}}, k \in \mathcal{K}. \tag{6.1}$$

This ensures that each UAV uses a given sub-channel for at most one communication direction (inbound or outbound), maintaining feasible and interference-aware spectrum usage.

6.2.3 U2U and G2U Transmission Rates

Without loss of generality, all UAVs are assumed to operate at a constant altitude H while collecting sensing data from the GUs. The trajectory of each UAV-i is represented as a sequence of spatial positions over the time horizon $\mathcal{T}$, denoted by $\mathcal{L}_i = [\boldsymbol{\ell}_i(t)]_{t \in \mathcal{T}}$. At any time slot t, the location $\boldsymbol{\ell}_i(t)$ is given by the three-dimensional coordinate $(x_i(t), y_i(t), z_i(t))$, where $z_i(t) = H$ for all t, indicating a constant flight height. The BS is located at the origin and its antenna height is denoted by H_b. Each UAV-i moves with a speed $v_i(t)$ that does not exceed the maximum allowable speed $v_{\max}$. The direction of motion at time t is specified by the unit vector $\boldsymbol{d}_i(t)$. The position of UAV-i in the next time slot is determined by the update rule $\boldsymbol{\ell}_i(t+1) = \boldsymbol{\ell}_i(t) + v_i(t)\boldsymbol{d}_i(t)$. The distance between UAV-i and UAV-j at time t is $d_{i,j}(t) = ||\boldsymbol{\ell}_i(t) - \boldsymbol{\ell}_j(t)||$.

Communications between UAVs and the BS typically occur via LoS links, which provide stable and high-quality channels. When UAV-i transmits sensing data to UAV-j over sub-channel $k \in \mathcal{K}$, the received signal power at UAV-j is given by $p_{j,i}^k(t) = p_i^k \beta_{i,j}^u (d_{i,j}(t))^{-\alpha_u}$. Here p_i^k denotes the transmit power of UAV-i on sub-channel k, and $\beta_{i,j}^u$ represents the constant power gain. The path loss is modeled as a function of the distance $d_{i,j}(t)$ between the two UAVs, raised to the power of $-\alpha_u$, where α_u is the path-loss exponent for U2U links. Since all UAVs share the same set of sub-channels, simultaneous transmissions on the same frequency can cause interference. If UAV-m ($m \neq i$) transmits to UAV-n over sub-channel k, the total interference to UAV-j is expressed as follows:

$$I_{j,i}^k(t) = \sum_{m \neq i} \sum_{n \neq j} \phi_{m,n}^k(t) p_m^k (d_{m,j}(t))^{-\alpha_u}. \tag{6.2}$$

This accounts for all active U2U links on the same sub-channel that contribute to interference at the receiver. Therefore, the total transmission rate over all sub-channels is obtained by summing the achievable rates on each sub-channel assigned to the link, given as follows:

$$o_{i,j}(t) = \sum_{k \in \mathcal{K}} \phi_{i,j}^k(t) \log \left(1 + \frac{p_{j,i}^k(t)}{\delta_k^2 + I_{j,i}^k(t)}\right), \tag{6.3}$$

where δ_k^2 denotes the noise power on sub-channel k. A similar channel model is applied to the links from GUs to UAVs. It is assumed that each UAV collects data only from GUs within its coverage area, and all such links are with LoS channel conditions due to the UAVs' elevated positions. Therefore, a similar log-distance model as that of U2U and U2B channels can be employed. Let q_m represent the GU-m's transmit power and $d_{m,i}(t)$ be the distance between GU-m and UAV-i. The data rate from GU-m to UAV-i is then given by

$$u_{i,m}(t) = \log \left(1 + q_m \beta_{m,i}^s (d_{m,i}(t))^{-\alpha_s}\right), \tag{6.4}$$

where α_s denotes the path-loss constant while $\beta_{m,i}^s$ represents the reference channel gain normalized by the noise power. Mutual interference among GUs is neglected in this model. This simplification is justified by the spatial separation of UAVs, which ensures that GUs transmitting to different UAVs do not significantly interfere with each another.

6.2.4 GUs' and UAVs' Data Queue Dynamics

Our primary goal is to jointly optimize the UAVs' trajectories and network formation strategy to minimize the transmission delay and total energy consumption. Each GU-m is initially equipped with a fixed volume of sensing data W_m that must be delivered to the BS. During each time slot, the sensing sub-slot t_s allows UAV-i to gather part of the data from GUs. In the subsequent offloading sub-slot t_o, the UAV transmits the buffered data to a neighboring UAV or directly to the BS. This process results in dynamic update of the GUs' and the UAVs' data queues over time:

$$W_m(t+1) = \left[W_m(t) - \sum_{i \in \mathcal{N}} x_{i,m}(t) s_{i,m}(t) \right]^+, \tag{6.5}$$

where $[X]^+ = \max\{0, X\}$. Here $s_{i,m}(t)$ represents the amount of sensing data collected from GU-m by UAV-i, and $x_{i,m}(t) \in \{0, 1\}$ indicates whether the UAV-i is associated with GU-m at time t. For instance, the GU-m's sensing data $s_{i,m}(t)$ will be collected by the UAV-i if $x_{i,m}(t) = 1$. A GU can connect to at most one UAV per slot, which is expressed as $\sum_{i \in \mathcal{N}} x_{i,m}(t) \leq 1$. Let $\mathcal{M}_i$ be the set of GUs within the coverage of UAV-i. The total sensing data collected by UAV-i at time t is given by

$$s_i(t) = \sum_{m \in \mathcal{M}_i} x_{i,m}(t) s_{i,m}(t). \tag{6.6}$$

In addition to newly collected sensing data $s_i(t)$, UAV-i may receive data from other UAVs. Let $\mathcal{I}_i(t) \triangleq s_i(t) + \sum_{j \neq i, j \in \mathcal{N}} o_{j,i}(t)$ represent the total incoming data to UAV-i's buffer. During the offloading sub-slot t_o, UAV-i forwards its buffered data either to a neighboring UAV or directly to the BS. The total out-going data from UAV-i is defined as $O_i(t)$, given as follows:

$$O_i(t) \triangleq o_{i,0}(t) + \sum_{j \neq i, j \in \mathcal{N}} o_{i,j}(t), \tag{6.7}$$

where $o_{i,j}(t)$ is the data rate defined in (6.3). The term $o_{i,0}(t)$ represents the data transmitted to the BS, while the summation accounts for data forwarded to other UAVs. Consequently, the data queue of UAV-i is updated as follows:

$$D_i(t+1) = \min\left\{[D_i(t) - O_i(t)]^+ + \mathcal{I}_i(t), D_{\max}\right\}. \tag{6.8}$$

In the case study, a simple data collection strategy is adopted, i.e., each UAV is associated with the GU that has the best signal quality within its coverage. Thus, the set $\mathcal{M}_i$ contains only one GU, denoted as GU-m, which exhibits the strongest channel to UAV-i. Given the sensing duration $t_s(t)$, the amount of sensing data collected is computed as

$$s_{i,m}(t) = t_s(t) \log\left(1 + q_m \beta^s_{m,i} (d_{m,i}(t))^{-\alpha_s}\right). \tag{6.9}$$

6.3 Joint Network Formation and Trajectory Planning

Following sensing data collection, each UAV-i transmits the buffered data to a neighboring UAV or the BS within its offloading sub-slot $t_{i,o}$. It then moves to the next location during the flying sub-slot $t_{i,f}$. The energy consumption of UAV-i is primarily influenced by its flying speed $v_i(t)$, hovering duration $t_{i,o}$, and flight time $t_{i,f}$. The operational energy $e_i(t)$ in time slot t is modeled using the well-known approach from [33]. The energy spent by UAV-i on data offloading is denoted by $p_i(t) = \sum_{j \in \mathcal{N}} \sum_{k \in \mathcal{K}} t_{i,o} p^k_{i,j} \phi^k_{i,j}(t)$. Hence, the energy minimization problem can be formulated as follows:

$$\min_{X,\Phi,\mathcal{L}_i,T} \sum_{i=1}^{N}\sum_{t=1}^{T}\left(e_i(t)+\sum_{j\in\mathcal{N}}\sum_{k\in\mathcal{K}} t_{i,o}p_{i,j}^k\phi_{i,j}^k(t)\right) \tag{6.10a}$$

$$s.t.\ (6.1)-(6.9), \tag{6.10b}$$

$$D_i(t)\le D_{\max} \text{ and } D_i(T)=0, \tag{6.10c}$$

$$W_m(0)=D_m \text{ and } W_m(T)=0, \tag{6.10d}$$

$$||\boldsymbol{\ell}_i(t+1)-\boldsymbol{\ell}_i(t)||\le v_{\max}(t)t_{i,f}, \tag{6.10e}$$

$$||\boldsymbol{\ell}_i(t)-\boldsymbol{\ell}_j(t)||\ge d_{\min}, \tag{6.10f}$$

$$\phi_{i,j}^k(t)\in\{0,1\} \text{ and } x_{i,m}(t)\in\{0,1\}, \tag{6.10g}$$

$$\forall i,j\in\mathcal{N}, \forall t\in\mathcal{T} \text{ and } \forall m\in\mathcal{M}, k\in\mathcal{K}. \tag{6.10h}$$

The goal is to jointly optimize the network formation $\Phi(t)$ and the binary association matrix $X(t)=[x_{i,m}(t)]_{i\in\mathcal{N},m\in\mathcal{M}}$ that defines the G2U connections in each time slot $t\in\mathcal{T}$. All these matrix variables should be jointly optimized with the UAVs' trajectories $\mathcal{L}_i$ for $i\in\mathcal{N}$. Additionally, the total number of time slots T required to complete data offloading from all GUs to the remote BS is also optimized. For simplicity, a fixed data collection strategy is adopted, i.e., each UAV-i only collects data from the GU with the strongest signal. Under this rule, the G2U association matrix $X(t)$ is determined solely by the UAVs' positions at each time slot. The constraints in (6.1)–(6.9) describe the sub-channel allocation and the buffer dynamics of both GUs and UAVs. The constraints in (6.10c)–(6.10d) guarantee that all sensing data from GUs is successfully delivered to the BS within T time slots. The inequalities in (6.10e) and (6.10f) impose limitations on UAVs' movements across time slots. In practice, the UAVs' transmit power in objective (6.10a) is negligible compared to the energy used for hovering and flying, so it can be omitted. Problem (6.10) is a mixed integer program and challenging to solve efficiently due to spatio-temporal couplings between the UAVs' network formation and trajectory planning. The joint optimization of time horizon T further complicates reformulation. When T is fixed, residual data may remain in the buffers of GUs or UAVs, potentially violating constraints (6.10c)–(6.10d) by the end of time horizon T. To address this, the objective in (6.10a) is modified by incorporating the remaining buffer data as penalty terms, and then reformulated into the following joint optimization:

$$\min_{\Phi,\mathcal{L}_i} \sum_{t=1}^{T}\left(\sum_{i=1}^{N}\left(\hat{e}_i(t)+\lambda_i D_i(t)\right)+\sum_{m=1}^{M} W_m(t)\right) \tag{6.11a}$$

$$s.t.\ (6.10b) \text{ and } (6.10e)-(6.10h), \tag{6.11b}$$

where λ_i is a parameter balancing the UAVs' energy consumption and data queue sizes. The UAVs' transmit power is omitted in (6.10a). For a fixed T, the objective is to jointly minimize the total energy consumption and the buffer queue lengths at both the UAVs and the GUs. In the following, we propose an approximate solution by decomposing problem (6.11) into two sub-problems: (i) UAV network formation and (ii) trajectory and resource planning, which are solved iteratively in a coordinated manner.

6.3.1 Adaptive Network Formation

In the first sub-problem, we adjust the UAVs' network formation $\Phi(t)$ based on their current trajectories $\mathcal{L}_i$. This allows us to simplify (6.11) as follows:

$$\min_{\Phi} \sum_{t=1}^{T}\left(\sum_{i=1}^{N}\left(\hat{e}_i(t) + \lambda_i D_i(t)\right) + \sum_{m=1}^{M} W_m(t)\right) \tag{6.12a}$$

$$s.t. \;\; \text{(6.10b) and (6.10g)} - \text{(6.10h)}. \tag{6.12b}$$

Problem (6.12) is a nonlinear integer program. Although solvable using existing branch-and-bound methods, its computational complexity is high due to the time-varying buffer states of both UAVs and GUs across time slots. Therefore, we introduce a lightweight heuristic algorithm, the energy- and delay-aware network formation (EDA-NF) algorithm, to update the network formation according to the UAVs' energy usage and buffer conditions. During each sub-slot $t_{i,o}$, UAV-i reports its current status to the BS, including its location $\boldsymbol{\ell}_i(t)$, current network formation $\Phi(t)$, energy consumption $\hat{e}_i(t)$, and buffer information $\lambda_i D_i(t) + \sum_{m\in\mathcal{M}_i} W_m(t)$, which reflects both its own buffer load $D_i(t)$ and the total traffic demands from GUs within its coverage.

After collecting status updates from all UAVs, the BS adjusts $\Phi(t)$ to balance energy consumption and queue lengths. We assume the UAVs' status information is small and introduces negligible overhead. To evaluate the UAV-i's traffic conditions with respect to the overall network traffic, we define a load balance coefficient $b_i(t)$ as follows:

$$b_i(t) = \frac{D_i(t)}{o_{i,0}(t)} - \frac{1}{N-1}\sum_{j\neq i, j\in\mathcal{N}} \frac{D_j(t)}{o_{j,0}(t)}, \tag{6.13}$$

which depends on UAVs' buffer levels and their U2B channel capacities. The term $\frac{D_i(t)}{o_{i,0}(t)}$ indicates the expected time delay for UAV-i to forward its buffered data to the BS through the U2B channel. The second term in (6.13) reflects the average time delay experienced by all other UAVs when transmitting via their U2B links. A large value of $|b_i(t)|$ is likely when there is a significant imbalance in traffic load between UAV-i and the rest, such as when UAV-i has a relatively large buffer occupancy but suffers from a weak U2B channel. In such a situation, UAV-i should avoid relying solely on the U2B channel and instead establish U2U connections with neighboring UAVs to offload part of its data through more favorable U2U channels. Note that

Algorithm 1 Energy- and Delay-Aware Network Formation (EDA-NF)

Initialize network formation $\phi_{i,0} = 1$ for $i \in \mathcal{N}$
Initialize (c_i, b_i) for all UAVs
Initialize $\mathcal{G}_1 \leftarrow \emptyset$, $\mathcal{G}_2 \leftarrow \mathcal{N}$
for $i \in \{1, 2, \ldots, N\}$ **do**
    Update $b_i(t)$ and $c_i(t)$
    $\mathcal{G}_1 \leftarrow \{i\}$ if $b_t > b_o$
    $\mathcal{G}_2 \leftarrow \{i\}$ if $b_t \leq b_o$
end for
Sort UAVs in $\mathcal{G}_1$ by the descent order of $c_i(t)$
Sort UAVs in $\mathcal{G}_2$ by the ascent order of $c_i(t)$
for any $i \in \mathcal{G}_1$ and $j \in \mathcal{G}_2$ **do**
    Check the distance $d_{i,j}$ between UAV-i and UAV-j
    If $d_{i,j} < d_k$ then $\phi_{i,j} \leftarrow 1$, $\phi_{i,0} \leftarrow 0$, $\phi_{j,0} \leftarrow 1$, and
    remove UAV-i from $\mathcal{G}_1$ and remove UAV-j from $\mathcal{G}_2$
end for
Return the network formation strategy $\Phi(t)$

frequent updates to the network formation are not necessary in every time slot. The network formation should remain unchanged if the load balance coefficients of all UAVs are sufficiently small.

In addition to the load balance coefficient $b_i(t)$, a cost function is defined for each UAV-i as follows:

$$c_i(t) = \hat{e}_i(t) + \lambda_i D_i(t) + \sum_{m=1}^{M} W_m(t), \tag{6.14}$$

which captures UAV-i's resource requirements, including its energy consumption $\hat{e}_i(t)$ and traffic load. A higher cost value $c_i(t)$ indicates that the UAV consumes more energy and experiences increased transmission delay due to poor channel conditions for data offloading. In contrast, the coefficient $b_i(t)$ reflects the quality of the direct U2B transmission link. We divide all UAVs into two groups according to the coefficient $b_i(t)$, represented as subsets $\mathcal{G}_1$ and $\mathcal{G}_2$. A larger $b_i(t)$ suggests that UAV-i has a relatively heavy workload and degraded channel condition. In such cases, we expect to establish a U2U link for UAV-i to offload data to a neighboring UAV with a lighter workload. Therefore, we apply a threshold-based classification between set $\mathcal{G}_1$ and $\mathcal{G}_2$. If $b_i(t)$ exceeds a threshold b_o, UAV-i is assigned to set $\mathcal{G}_1$. Otherwise, it is placed in set $\mathcal{G}_2$. UAVs in set $\mathcal{G}_1$ are allowed to establish U2U links with other UAVs, whereas UAVs in set $\mathcal{G}_2$ maintain direct U2B connections.

The core idea of the EDA-NF algorithm is to establish U2U connections between UAVs from the two subsets, enabling UAVs in $\mathcal{G}_1$ with heavy loads to offload data to those in $\mathcal{G}_2$ with lighter loads. For the UAV-i in subset $\mathcal{G}_1$ that has the highest cost value $c_i(t)$, a U2U channel can be assigned. The UAV-j in subset $\mathcal{G}_2$ with the smallest cost value $c_j(t)$ is selected as the relay node for UAV-i. It is necessary to ensure that the U2U channel from UAV-i to UAV-j satisfies a minimum required data rate.

The complete steps are summarized in Algorithm 1. Network state information is typically required in Algorithm 1 to compute the values in (6.13) and (6.14). This can be collected by the UAVs and sent to the BS together with their data transmissions. The threshold b_o can be determined experimentally during an offline phase.

6.3.2 MADRL for Trajectory Optimization

With the network formation $\Phi(t)$ fixed, the second sub-problem focuses on optimizing the UAVs' trajectories as follows:

$$\min_{\mathcal{L}_i} \sum_{t=1}^{T}\left(\sum_{i=1}^{N}\left(\hat{e}_i(t)+\lambda_i D_i(t)\right)+\sum_{m=1}^{M} W_m(t)\right) \tag{6.15a}$$

$$s.t.\ \ (6.10b) \text{ and } (6.10e)-(6.10f). \tag{6.15b}$$

This trajectory optimization problem is inherently high-dimensional and challenging to solve directly due to the coupled spatial and temporal dependencies among UAVs. To address this complexity, we adopt a model-free DRL approach by casting problem (6.15) as a MDP, defined by the tuple $(\mathcal{S}, \mathcal{A}, \mathcal{R})$. Here, $\mathcal{S}$ denotes the global state space, $\mathcal{A}$ represents the joint action space, and $\mathcal{R}$ is the reward function that evaluates the quality of each state-action transition. In this MDP formulation, the global state $\mathbf{s}(t)$ represents the local observations of all UAVs, expressed as $\mathbf{s}(t) = (\mathbf{s}_1(t), \mathbf{s}_2(t), \ldots, \mathbf{s}_N(t))$. The local state $\mathbf{s}_i(t)$ for UAV-i includes its current position $\boldsymbol{\ell}_i(t)$, its connectivity status $\{\phi_{i,j}(t)\}_{j\in\tilde{\mathcal{N}}}$, remaining energy $E_i(t)$, and buffered data volume $D_i(t)$. Similarly, the joint action $\mathbf{a}(t)$ is composed of individual actions, i.e., $\mathbf{a}(t) = (\mathbf{a}_1(t), \mathbf{a}_2(t), \ldots, \mathbf{a}_N(t))$, where each $\mathbf{a}_i(t)$ consists of the flying direction $\mathbf{d}_i(t)$ and speed $v_i(t)$ chosen at time t.

When UAV-i executes action $\mathbf{a}_i(t)$ in state $\mathbf{s}(t)$, it receives an immediate reward $R_i(\mathbf{s}(t), \mathbf{a}_i(t))$, which is influenced not only by its own decision but also by the concurrent actions $\mathbf{a}_{-i}(t)$ of other UAVs. This reward is decomposed into four components including energy efficiency, transmission performance, sensing contribution, and safety compliance. The energy-related reward penalizes high consumption and is defined as:

$$R_{i,e}(t) = -\hat{e}_i(t), \tag{6.16}$$

encouraging UAV-i to minimize propulsion and operational energy usage. To incentivize timely data delivery, a transmission reward is granted based on the total data volume successfully forwarded to the BS or neighboring UAVs:

$$R_{i,d}(t) = o_{i,0}(t) + \sum_{j\in\mathcal{N}} o_{i,j}(t). \tag{6.17}$$

A sensing reward is introduced to motivate UAVs to actively gather data from GUs within their coverage area. For UAV-i, this is quantified as the sum of collected sensing data:

$$R_{i,s}(t) = \sum_{m \in \mathcal{M}_i} s_{i,m}(t). \tag{6.18}$$

Additionally, a penalty term $R_{i,p}(t)$ is imposed to enforce the minimum safety distance constraint in (6.10f). This penalty is activated whenever UAV-i comes too close to another UAV:

$$R_{i,p}(t) = \mu \sum_{j \in \mathcal{N}} \mathbf{I}(||\boldsymbol{\ell}_i(t) - \boldsymbol{\ell}_j(t)|| < d_{\min}), \tag{6.19}$$

where $\mathbf{I}(\cdot)$ is an indicator function that returns 1 if the condition is true, and 0 otherwise. Finally, the total reward for UAV-i is a weighted combination of these components:

$$R_i(\mathbf{s}(t), \mathbf{a}(t)) = \gamma_1 R_{i,e}(t) + \gamma_2 R_{i,d}(t) + \gamma_3 R_{i,s}(t) - R_{i,p}(t), \tag{6.20}$$

where γ_1, γ_2, and γ_3 are positive weights that balance the importance of energy savings, data delivery, and sensing coverage in the learning process.

In multi-UAV-assisted wireless networks, the observation of each UAV is influenced not only by its own actions but also by the behaviors of other UAVs. We employ the MADDPG algorithm to effectively learn optimal trajectories under such interdependencies, leveraging a centralized training with decentralized execution paradigm [34]. Each UAV-i operates as an autonomous agent governed by a policy parameterized by $\boldsymbol{\theta}_i$, producing its action $\mathbf{a}_i$ through a deterministic policy $\pi_i(\mathbf{o}_i|\boldsymbol{\theta}_i)$ based on its local observation $\mathbf{o}_i$. It is important to note that $\mathbf{o}_i$ may not fully capture the global system state due to partial observability inherent in distributed settings [13]. Each UAV-i maintains a Q-value function $Q_i^{\pi}(\mathbf{o}, \mathbf{a}|\boldsymbol{w}_i)$, which estimates the long-term return and depends on the joint observations and actions of all agents. Denoting the observations and actions of all other UAVs as $\mathbf{o}_{-i}$ and $\mathbf{a}_{-i}$, respectively, the policy gradient for UAV-i can be reformulated as:

$$\nabla_{\theta_i} J(\boldsymbol{\theta}_i) = \mathbb{E}_{\mathcal{B}}[\nabla_{\mathbf{a}_i} Q_i^{\pi}(\mathbf{o}, \mathbf{a}|\boldsymbol{w}_i)\nabla_{\theta_i}\pi_i(\mathbf{o}_i|\boldsymbol{\theta}_i)], \tag{6.21}$$

where $\mathbf{o} \triangleq (\mathbf{o}_i, \mathbf{o}_{-i})$ and $\mathbf{a} \triangleq (\mathbf{a}_i, \mathbf{a}_{-i})$ represent the concatenated observations and actions of all UAVs. The expectation is taken over transitions stored in the experience replay buffer $\mathcal{B}$. As shown in (6.21), while the actor network $\pi_i(\mathbf{o}_i|\boldsymbol{\theta}_i)$ can be updated using only local information, the critic network $Q_i^{\pi}(\mathbf{o}, \mathbf{a}|\boldsymbol{w}_i)$ requires access to global system information during training. This structural separation enables the use of centralized learning, i.e., during the offline phase, the BS gathers observations and actions from all UAVs to train the critic networks with full system knowledge.

Once training is complete, the learned actor networks are deployed to individual UAVs, allowing them to make decisions autonomously without requiring continuous global coordination. With a fixed network formation $\Phi(t)$, each UAV-i determines its next movement by applying the MADDPG algorithm to select the optimal flying action and update its trajectory $\mathcal{L}_i$. As UAVs move along their paths, they collect sensing data from associated GUs and forward it either directly to the BS or via intermediate UAVs. Concurrently, each UAV periodically reports its operational status to the BS. Based on these updates, the BS evaluates the current network structure by computing the load balance coefficients and cost values defined in (6.13) and (6.14). If significant imbalances in energy usage or data backlog are detected, the BS triggers a network reformation to redistribute traffic load and restore equilibrium of the UAV-assisted wireless network.

6.4 Bayesian Optimization Enhanced MADDPG

While MADDPG offers a powerful framework for addressing complex control tasks in multi-agent networks, its direct application to multi-UAV's trajectory optimization faces several practical challenges. A primary issue arises from the need to aggregate observations from all UAVs to enable joint action selection during training. Each UAV must transmit its local state, such as channel conditions, energy levels, and buffer queue lengths, to the BS. In dynamic networks with time-varying topologies, this process is hindered by delays inherent in multi-hop relay transmissions. Consequently, the global information available at the BS for centralized learning may quickly become outdated due to the UAVs' continuous mobility, leading to inaccurate policy updates. Moreover, the dimensionality of both the state and action spaces grows significantly with the number of UAVs, exacerbating the complexity of the learning problem. The communication burden of sharing local observations and actions across the network also escalates. This not only prolongs the training process but also destabilizes learning convergence, making scalable and real-time deployment difficult.

6.4.1 Action Estimation via Bayesian Optimization

To enhance the learning efficiency of MADDPG, we integrate Bayesian optimization (BO) for estimating optimal flying actions for each UAV in the subsequent time step [35]. Given UAVs' historical trajectory observations, BO offers a data-driven approach to predict high-performing actions by leveraging past sampling data. Unlike purely random exploration, BO-driven action selection provides a more informed direction for policy search, effectively steering trajectory learning toward higher-reward regions of the policy space. This mechanism can be interpreted as a form of semi-supervised learning, where the BO module supplies actionable priors to the DRL agent, significantly improving sample efficiency, particularly during the initial learning phase, where conventional exploration strategies often require extensive trial-and-error to achieve meaningful performance.

The proposed framework, referred to as BO-MADDPG, consists of three key components, as illustrated in Fig. 6.1. First, the action estimation module receives the most recent trajectory data from each UAV and employs BO to predict the most favorable flying positions for the next time slot. This estimated action is then fed into the critic network alongside the action generated by the actor network. The critic evaluates the expected return of both actions under the current system state, enabling a comparative assessment. Based on this evaluation, the system probabilistically selects the action with a higher reward, thereby avoiding unproductive exploratory moves. This selective action execution effectively narrows the action space and accelerates convergence. While model-free MADDPG relies on trial-and-error to refine UAVs' trajectories, BO offers a more direct and computationally lightweight method for action exploration. This allows us to decouple the multi-UAV's joint trajectory planning into N independent single-UAV's planning subproblems, each solved using only local observations. By applying BO individually to each UAV, we generate action estimates that serve as intelligent guidance for the respective DRL agents. These estimates help agents focus exploration on promising regions of the action space, leading to faster learning and improved overall policy performance.

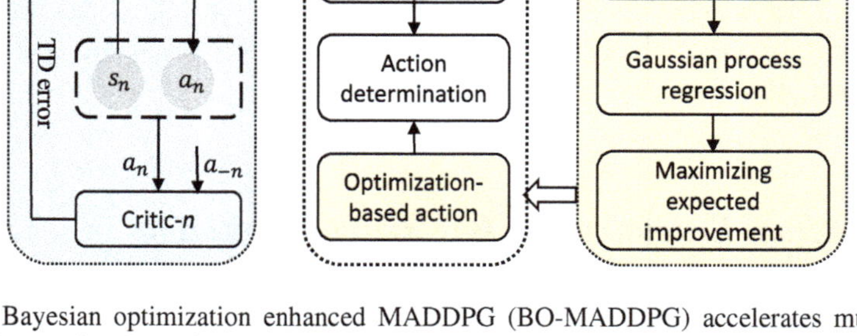

Fig. 6.1 Bayesian optimization enhanced MADDPG (BO-MADDPG) accelerates multi-agent DRL by integrating Bayesian optimization (BO) to guide UAVs' action selection. BO predicts high-reward actions using historical trajectories, providing informed priors that reduce random exploration. These estimates are evaluated by the critic, enabling selective execution. This semi-supervised approach improves sample efficiency and accelerates convergence in UAVs' trajectory learning

6.4.2 Data-Driven Probabilistic Model

For each UAV-i, let $\boldsymbol{\ell}_i(t)$ represent its position at time t, and denote $s_i(t) = \sum_{m\in\mathcal{M}_i} s_{i,m}(t)$ as the total sensing data during the t-th time slot, where $s_{i,m}(t)$ is defined in (6.9). We introduce a mapping function:

$$f_i : \boldsymbol{\ell}_i(t) \rightarrow s_i(t) + \epsilon_i(t) \tag{6.22}$$

that associates each location $\boldsymbol{\ell}_i(t)$ with the observed data volume $s_i(t)$, perturbed by an error term $\epsilon_i(t)$. This term accounts for modeling inaccuracies and environmental uncertainty, assumed to follow a zero-mean Gaussian distribution and to be independently and identically distributed.

Based on historical observations, BO enables us to construct a probabilistic surrogate model $\mathcal{P}(f_i(\mathcal{H}_t)|\mathcal{D}_i(t))$ that can be used to predict data distribution over locations. Here the historical sampling points $\mathcal{D}_i(t) \triangleq (\boldsymbol{\ell}_i(\tau), s_i(\tau))_{\tau\in\mathcal{H}_t}$ contain past location-data pairs, where $\mathcal{H}_t = \{t - t_o, \ldots, t-1, t\}$ indexes the recent time window. The model $\mathcal{P}(f_i(\mathcal{H}_t)|\mathcal{D}_i(t))$ captures the posterior distribution of the latent function values $f_i(\mathcal{H}_t) \triangleq \{f_i(\tau)\}_{\tau\in\mathcal{H}_t}$ corresponding to the trajectory points $\{\boldsymbol{\ell}_i(\tau)\}_{\tau\in\mathcal{H}_t}$. As the number of historical samples $|\mathcal{D}_i(t)|$ increases, the posterior becomes more informative, leading to more accurate predictions. According to Bayes' rule, the posterior is proportional to the product of the likelihood and the prior:

$$f_i(\mathcal{H}_t)|\mathcal{D}_i(t) \sim \mathcal{P}(\mathcal{D}_i(t)|f_i(\mathcal{H}_t))\mathcal{P}(f_i(\mathcal{H}_t)). \tag{6.23}$$

Due to the lack of prior knowledge about GUs' distributions and traffic patterns, we adopt a multivariate Gaussian process as a flexible prior model [35]:

$$f_i(\mathcal{H}_t) \sim \mathcal{G}(\boldsymbol{\mu}_i(\mathcal{H}_t), \boldsymbol{V}_i(\mathcal{H}_t)), \tag{6.24}$$

where $\boldsymbol{\mu}_i(\mathcal{H}_t)$ is the mean vector and $\boldsymbol{V}_i(\mathcal{H}_t) \triangleq \{v_{\tau,\tau'}\}_{\tau,\tau'\in\mathcal{H}_t}$ is the covariance matrix (or kernel), encoding the similarity between observations at different time steps. Initially we set $\boldsymbol{\mu}_i = \boldsymbol{0}$. The kernel element $v_{\tau,\tau'}$ is defined as:

$$v_{\tau,\tau'}(\mathcal{H}_t) = \exp\left(-\frac{1}{2}||\boldsymbol{\ell}_i(\tau) - \boldsymbol{\ell}_i(\tau')||^2\right), \tag{6.25}$$

which assigns higher correlation values to trajectory points that are spatially closer. Given this Gaussian prior (6.24) and the assumption of Gaussian likelihood $\mathcal{P}(\mathcal{D}_i(t)|f_i(\mathcal{H}_t))$, the posterior distribution $\mathcal{P}(f_i(\mathcal{H}_t)|\mathcal{D}_i(t))$ in (6.23) is also Gaussian, with analytically computable mean and variance [36], enabling efficient Bayesian inference and action prediction.

6.4.3 Evaluating Next Trajectory Point

We now focus on predicting the value of the objective function at a candidate next location $\boldsymbol{\ell}_i(t+1)$ for UAV-i, denoted as $f_i(t+1)$. Conditioning on the historical dataset $\mathcal{D}_i(t)$, the posterior distribution of $f_i(t+1)$ is updated and can be expressed as follows:

$$f_i(t+1)|\mathcal{D}_i(t) \sim \mathcal{G}(\mu_i(t+1), \sigma_i^2(t+1)), \tag{6.26}$$

where the updated mean $\mu_i(t+1)$ and variance $\sigma_i^2(t+1)$ are derived from the prior parameters $(\boldsymbol{\mu}_i(\mathcal{H}_t), \boldsymbol{V}_i(\mathcal{H}_t))$ through Gaussian process inference. Further details on the derivation can be found in [35] and Sect. 2.1 of [36]. As the number of historical samples $|\mathcal{D}_i(t)|$ grows, the posterior distribution in (6.26) becomes increasingly concentrated around the true function value, enhancing the reliability of predictions. This probabilistic model enables UAV-i to assess potential future locations in terms of expected data collection performance. To maximize the amount of data gathered from GUs, the UAV selects the next waypoint $\boldsymbol{\ell}_i(t+1)$ that maximizes the expected improvement in sensing gain. We define the expected improvement function $z_{i,t}(\boldsymbol{\ell}_i)$ as the expected value of exceeding the current best observation:

$$z_{i,t}(\boldsymbol{\ell}_i) = \mathbb{E}[\max\{0, f_i(\boldsymbol{\ell}_i) - f_i^*(\mathcal{D}_i(t))\}], \tag{6.27}$$

where $f_i^*(\mathcal{D}_i(t)) = \max_{\boldsymbol{\ell}_i \in \mathcal{D}_i(t)} f_i(\boldsymbol{\ell}_i)$ represents the highest function value observed so far. The optimal next location is then obtained by maximizing the acquisition function as follows:

$$\boldsymbol{\ell}_i(t+1) = \arg\max z_{i,t}(\boldsymbol{\ell}_i). \tag{6.28}$$

In practice, this optimization is subject to physical constraints such as maximum speed and flight range of the UAV. In contrast to conventional model-free MADDPG, this Bayesian approach provides a directed and data-efficient exploration mechanism, guiding the UAV's action exploration toward regions with higher potential data collection performance. Upon reaching the new location, the UAV collects updated sensing data from associated GUs and reports the outcome to the BS, which relies on this information to assess the effectiveness of the current network formation strategy.

6.4.4 Action Estimation for MADDPG

The BO-MADDPG algorithm is outlined in Algorithm 2. At each time step, each UAV-i first uses its actor network to generate a candidate flying action $\mathbf{a}_i$ based on its local observation. Concurrently, each UAV's BO module computes the posterior distribution $\mathcal{P}(f_i(t+1)|\mathcal{D}_i(t))$ using its historical data $\mathcal{D}_i(t)$. Leveraging this probabilistic model, the next potential location $\boldsymbol{\ell}_i(t+1)$ is determined by maximizing the expected improvement defined in (6.27), thereby guiding the UAV toward positions

Algorithm 2 BO-MADDPG for UAVs' Trajectory Planning

1: Initialize network formation $\Phi(t)$ and trajectories $\mathcal{L}_i(t)$
2: Initialize observations $\mathcal{D}_i$ for all UAVs
3: **for** Episode = $\{1, 2, \ldots, \text{MAX} = 200K\}$ **do**
4:   Each UAV constructs its state $\mathbf{s}_i(t)$.
5:   **for** each UAV agent $i \in \{1, 2, \ldots, N\}$ **do**
6:     Estimate posterior $f_i(t+1)$ based on sampling points $\mathcal{D}_i(t)$ using Bayesian optimization
7:     Maximize the expected improvement in (6.27)
8:     Obtain the next trajectory point $\boldsymbol{\ell}_i(t+1)$
9:     Convert $\boldsymbol{\ell}_i(t+1)$ into the UAV's action $\hat{\mathbf{a}}_i(t)$
10:     Actor network updates action $\mathbf{a}_i(t) = \pi_i(\mathbf{o}_i|\boldsymbol{\theta}_i)$
11:     Evaluate the action-value $(q_i, \hat{q}_i)$ for $(\mathbf{a}_i, \hat{\mathbf{a}}_i)$
12:     Take the action $\mathbf{a}_i^*$ with a higher action-value
13:     Record state transition $s_i(t+1)$ and the reward $R_i(t)$
14:     Store the transition and update the history $\mathcal{D}_i(t)$
15:     Update the UAV's actor- and critic-network
16:   **end for**
17:   $t \leftarrow t+1$
18:   Update $b_i(t)$ and $c_i(t)$ for all UAVs
19:   Update network formation $\Phi(t)$ by Algorithm 1
20: **end for**

with high data collection potential. This estimated action is denoted as $\hat{\mathbf{a}}_i(t)$, and all UAVs' action estimates form the joint action vector $\{\hat{\mathbf{a}}_1(t), \hat{\mathbf{a}}_2(t), \ldots, \hat{\mathbf{a}}_N(t)\}$. For each UAV-$i$, the critic network evaluates both the action $\mathbf{a}_i$ from its actor-network and the BO-driven action $\hat{\mathbf{a}}_i(t)$, producing corresponding Q-values q_i and $\hat{q}_i$. The value $\hat{q}_i$ is computed by assessing the predicted return of the BO-driven location $\boldsymbol{\ell}_i(t+1)$. By comparing q_i and $\hat{q}_i$, the system selects a more favorable action for execution. Specifically, the final action $\mathbf{a}_i^*$ for UAV-i is chosen as $\hat{\mathbf{a}}_i(t)$ if $\hat{q}_i > q_i$, otherwise the actor's output $\mathbf{a}_i$ is retained. This selection mechanism allows the UAV to benefit from both DRL-based policy learning and data-driven exploration, resulting in a more efficient trajectory update.

The proposed framework operates in two stages, i.e., adaptive network formation governed by each UAV's cost parameters $(b_i(t), c_i(t))$, and a BO-driven multi-agent trajectory planning module. In Algorithm 2, lines 6–10 describe how each UAV-i determines its next position using both the actor-network and the BO module. Given the pair of candidate actions $(\mathbf{a}_i, \hat{\mathbf{a}}_i)$ and their corresponding Q-value estimates $(q_i, \hat{q}_i)$, UAV-i adopts a greedy selection strategy, choosing the action associated with the higher estimated Q-value. The selected action $\mathbf{a}_i^*$ is then executed, and the resulting state transition is recorded, as detailed in lines 11–15. Following trajectory execution, each UAV-i assesses its current traffic load and cost metrics, triggering an update to the network formation using Algorithm 1. For practical deployment, computational tasks can be distributed between the BS and individual UAVs. The BS may handle system initialization and pre-train the actor and critic networks centrally. Subsequently, each UAV operates autonomously, generating local actions from its own observations and optionally invoking the BO module to refine trajectory decisions. To reduce onboard computational load and enhance practical feasibility, BO can be selectively executed, rather than at every time step. This intermittent application on demand maintains effective guidance for action exploration while minimizing resource usage, striking a balance between performance gains and computational feasibility in real-world deployment scenarios.

6.5 Numerical Evaluation

This section presents simulation results to evaluate the effectiveness of the proposed joint network formation and trajectory optimization scheme for multi-UAV-assisted wireless networks. GUs are randomly deployed within a $2\times2\,\text{km}^2$ rectangular area, with their x-y coordinates normalized to the interval $[-1, 1]$ for computational convenience. The simulation environment includes 3 sub-channels, each with 1 MHz bandwidth, operating at 2 GHz carrier frequency with a path-loss exponent of 2. The BS is positioned outside the main service region, specifically in the upper right corner of the rectangular area. Both UAVs and GUs operate with a constant transmit power of 23 dBm. The UAVs are set to fly at a fixed altitude of 100 m with a maximum speed of 20 m/s, while each GU has 10 Mbits of data to transmit under a noise power of -90 dBm. Each training episode of the proposed learning algorithm consists of 200 cycles with a batch size of 256, and the actor and critic networks use learning rates of 10^{-3} and 10^{-4} respectively. Action exploration is controlled by an ϵ-greedy parameter and noise rate, both set to 0.1, to balance exploration and exploitation during training. These simulation parameters are set similar to the configurations in [8] for consistency and comparative analysis.

6.5.1 Bayesian Optimization Improves Trajectory Planning

For a comprehensive performance analysis, we first introduce the Layered-MADDPG framework, which combines the adaptive network formation algorithm with model-free MADDPG for trajectory learning. This separation reduces the dimensionality of the action space, facilitating more efficient exploration. Building upon this, we enhance the framework by integrating a BO module to model the sensing environment and guide policy exploration, leading to the proposed BO-MADDPG framework, which significantly boosts learning efficiency. The UAVs' trajectories generated by different strategies, including BO alone, Layered-MADDPG, and BO-MADDPG, are illustrated in Fig. 6.2. Each UAV's trajectory is represented by a distinct color and starting from a random initial position. Figure 6.2a depicts trajectories planned based on BO method. In this decentralized approach, each UAV independently selects its next position based on its own past observations, without coordination via centralized training. Consequently, the absence of inter-agent cooperation limits overall performance in a multi-UAV network. This lack of coordination often results in overlapping service areas, as seen in Fig. 6.2a, leading to redundant coverage and inefficient resource usage.

In contrast, Fig. 6.2b shows the UAVs' trajectories generated by the Layered-MADDPG algorithm, which operates under a centralized training and decentralized execution paradigm. During training, global state information is utilized to learn cooperative policies. As a result, the UAVs learn to partition the service area into distinct regions, assigning each to a specific UAV. The GUs are grouped into three spatially separated clusters and each UAV focuses on data collection within its designated zone. This task decomposition enhances cooperation and can reduce overall energy consumption and transmission latency. As shown in Fig. 6.2c, BO-MADDPG also achieves a clustered division of GUs. However, it further improves performance by prioritizing GUs located at cluster boundaries (e.g., GU-7), allocating higher transmission rates to ensure reliable connectivity. Moreover, the UAVs' trajectories generated by BO-MADDPG exhibit greater consistency compared to Layered-MADDPG. In Fig. 6.2d, we evaluate BO-MADDPG in a different case as all UAVs' trajectories begin from the same starting point. It enables the UAVs to effectively disperse and cover complementary areas, demonstrating its strong capability in fostering cooperative behavior.

Figure 6.3 presents the amount of sensing data collected by each UAV over time, where $R_{i,s}(t)$ denotes the data gathered by UAV-i at time t, as defined in (6.18). The results in Fig. 6.3a show that BO-MADDPG consistently collects more sensing data than the other methods, validating its effectiveness of combining model-free DRL with informative guidance from partial environmental modeling. A similar trend is observed in Fig. 6.3b, corresponding to the scenario in Fig. 6.2d where UAVs start from a shared location. In both cases, BO-MADDPG outperforms the other baselines in sensing data collection. Both Layered-MADDPG and BO-MADDPG adopt a hierarchical design that decouples network formation from trajectory planning. For comparison, we also evaluate the conventional model-free MADDPG approach that simultaneously optimizes UAVs' mobility and network formation strategies, denoted as Joint-MADDPG in Fig. 6.4. Thus, the action $\mathbf{a}_i(t)$ includes flying direction $\mathbf{d}_i(t)$, speed $v_i(t)$, and network formation variables $\phi_{i,j}(t)$. The learning convergence of these algorithms is shown in Fig. 6.4a. BO-MADDPG converges faster and attains a higher reward, with a sharp initial increase and a smoother, more stable learning curve compared to Layered-MADDPG and Joint-MADDPG. In contrast, Joint-MADDPG requires more iterations to develop an effective policy. Figure 6.4b compares the reward variance during training, where lower values indicate more stable learning. BO-MADDPG quickly stabilizes with minimal fluctuations, while Joint-MADDPG exhibits high variance due to unguided exploration.

6.5.2 Trajectory-Adaptive Network Formation

This section evaluates the UAVs' trajectory planning under the energy- and delay-aware adaptive network formation strategy, as outlined in Algorithm 1. Figure 6.5a

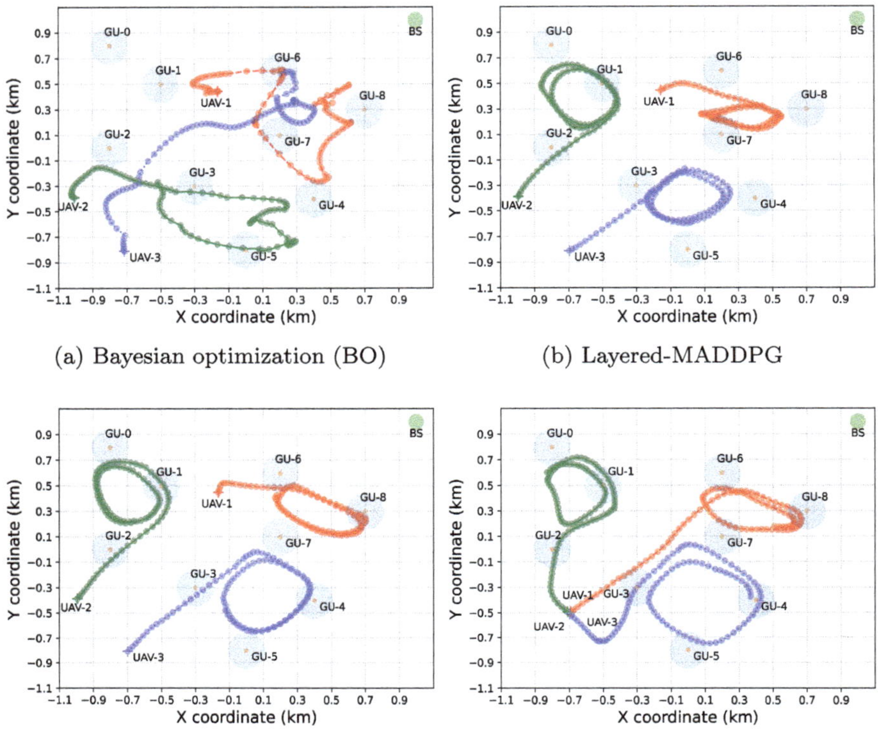

(a) Bayesian optimization (BO)

(b) Layered-MADDPG

(c) BO-MADDPG (Case I)

(d) BO-MADDPG (Case II)

Fig. 6.2 The comparison of BO, Layered-MADDPG, and BO-MADDPG for UAVs' trajectory planning. BO alone lacks coordination, causing redundant coverage. Layered-MADDPG enables cooperative area partitioning via centralized training. BO-MADDPG further improves performance by prioritizing boundary users and enhancing trajectory consistency, demonstrating superior cooperation and coverage efficiency, even from the same starting positions

displays three UAVs's trajectories and highlights the dynamic evolution of their communication links. Different markers distinguish between U2U and U2B transmission modes. Hollow circles indicate direct U2B communication, while other markers denote U2U relaying. As a UAV moves farther from the BS, direct communication becomes less reliable, prompting a shift to U2U relaying through neighboring UAVs to maintain high data rates and lower latency. Conversely, when a UAV approaches the BS and its U2B channel improves, it switches back to direct U2B transmission. In Fig. 6.5a, UAV-1 operates in proximity to the BS, whereas UAV-2 and UAV-3 serve distant regions. Consequently, UAV-1 frequently acts as a relay, receiving data from UAV-2 or UAV-3 via high-speed U2U links and forwarding it to the BS. When UAV-3 moves closer to the BS and its U2B link quality improves, it terminates the U2U connection and begins direct U2B data forwarding. Additionally, UAV-3 can serve as a relay for UAV-2 when UAV-1 is overloaded, demonstrating the flexibility of the adaptive formation. Focusing on specific time intervals, during steps 0–8 in

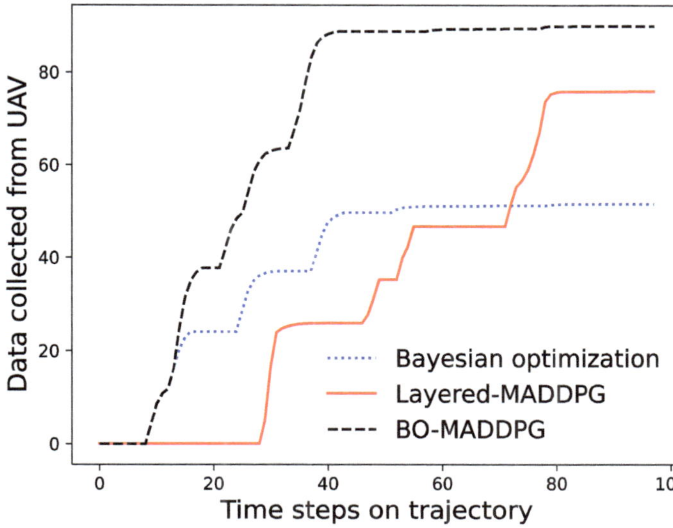

(a) BO-MADDPG (Case I)

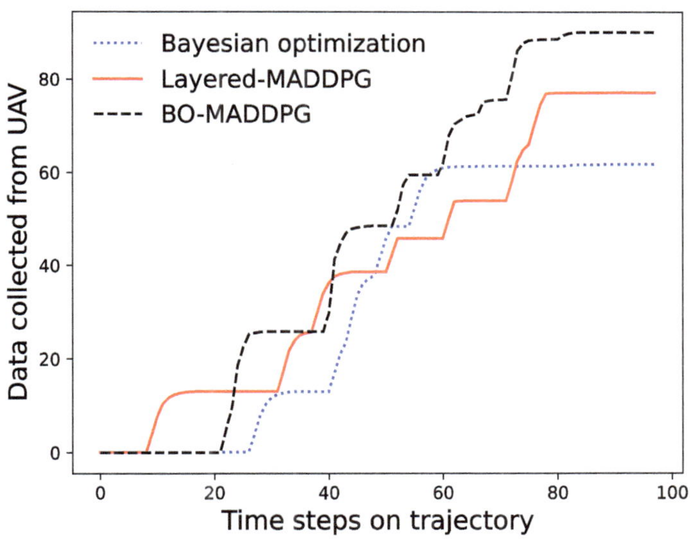

(b) BO-MADDPG (Case II)

Fig. 6.3 BO-MADDPG outperforms baselines in sensing data collection, validating the effectiveness of the hybrid approach under various deployment scenarios. It consistently gathers more data over time, even when UAVs start from the same location

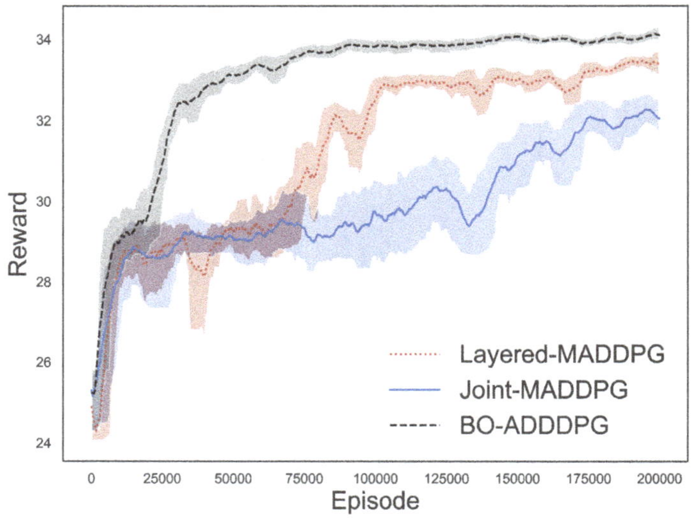

(a) Better reward performance

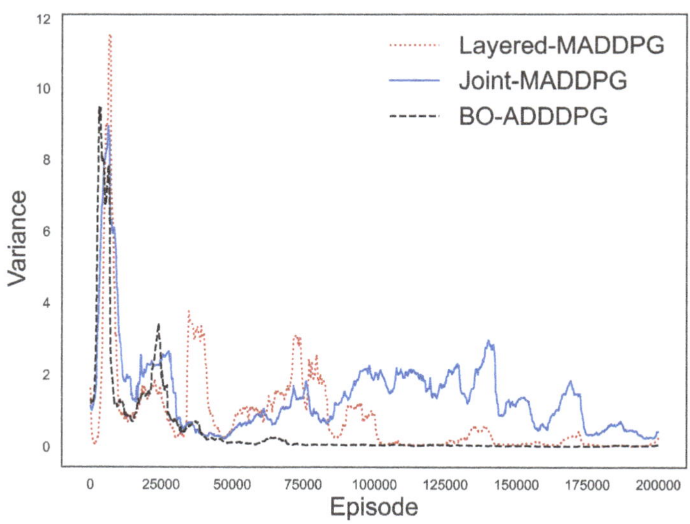

(b) More stable learning

Fig. 6.4 BO-MADDPG achieves smoother learning, faster convergence, and higher rewards than Layered-MADDPG and Joint-MADDPG, thanks to its hierarchical design and BO-driven exploration. It reduces action space by decoupling network formation from trajectory planning. Joint-MADDPG jointly optimizes all variables, suffering from slower learning and high variance due to unguided exploration in a larger policy space

Fig. 6.5a, none of the UAVs are within sensing range of any GU. From steps 9 to 13, UAV-2 begins collecting data from GUs, while UAV-1 and UAV-3 remain out of sensing range. During this phase, UAV-3 relays UAV-2's data to the BS via a U2U link. In the steps 14–20, UAV-2 switches from U2U to direct U2B communication as its link to the BS improves, while UAV-3 establishes a U2U connection with UAV-1 for relaying. UAV-1, being closest to the BS, can rapidly offload its own and buffered data through its favorable U2B channel. Similar dynamics govern the subsequent phases of the mission.

Figure 6.5b depicts the variation of each UAV's buffer occupancy throughout the flight. Initially, UAV-1 maintains a minimal buffer level due to its strong U2B channel, enabling prompt data offloading. In contrast, UAV-2 and UAV-3 experience limited U2B rates, as they are far from the BS, causing an sharp increase in their buffers. As their buffers grow, their respective cost functions increase, triggering a network reconfiguration. This prompts the establishment of U2U links between UAV-2/3 and UAV-1. Consequently, as seen in Fig. 6.5b, UAV-1's buffer size rises due to incoming relay data, while the buffers of UAV-2 and UAV-3 decrease. Through this cooperative offloading mechanism, all UAVs successfully empty their buffers simultaneously, minimizing mission duration and balancing UAVs' resource utilization.

6.5.3 Comparing Different Network Formation Schemes

This section presents a performance evaluation of the adaptive network formation strategy in Algorithm 1, demonstrating its effectiveness in enhancing the system reward compared to several benchmark approaches. The first baseline is a non-cooperative scheme, widely adopted in existing studies, where each UAV relies solely on direct U2B links to offload data to the BS, without inter-UAV coordination. The second baseline mimics a multi-hop relay network as in [9]. To incentivize relaying, we implement a buffer-driven mechanism, i.e., UAVs establish U2U connections with neighbors when their onboard buffer occupancy exceeds a predefined threshold. Buffer backlog is a standard metric in multi-hop networks for routing and congestion control [37]. The third baseline is the dynamic network formation (dynamic-NF) method in [38], which computes a cost metric for each UAV based on its buffer load, residual data volume, and energy usage. UAVs dynamically adjust their U2U links by comparing their cost with that of immediate neighbors. In contrast, Algorithm 1 introduces an additional load balance coefficient to reflect each UAV's transmission capacity, beyond just cost-based decisions.

Figure 6.6a compares the convergence behavior and final reward of these strategies. The non-cooperative approach converges most rapidly due to its simplicity, but achieves the lowest cumulative reward, as it ignores cooperative gains. The buffer-based method incorporates cooperation by reacting to buffer congestion, but it overlooks other critical factors such as energy constraints and GU workload distribution, which can influence optimal trajectory decisions. The dynamic-NF strategy achieves a higher reward than the first two baselines by considering a broader set of system

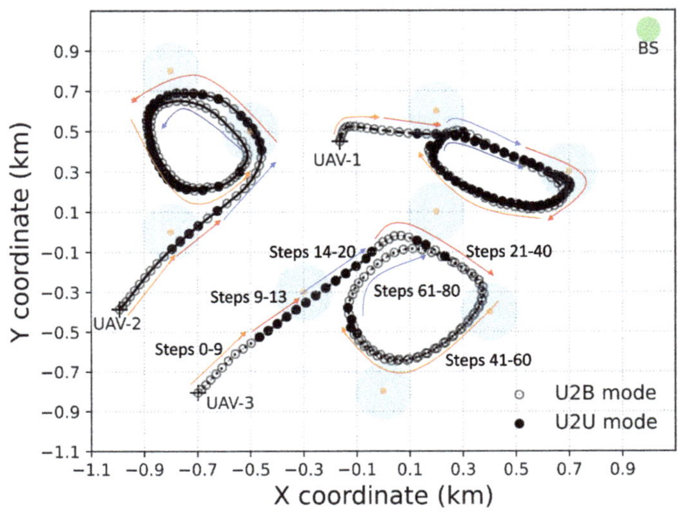

(a) Adaptive network formation

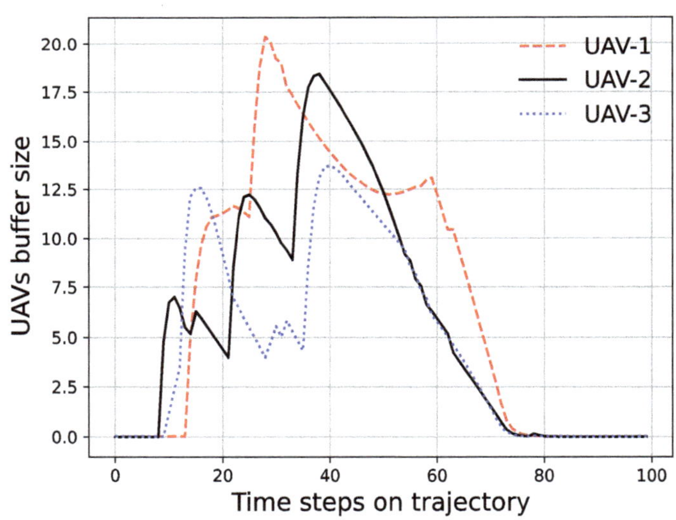

(b) The UAVs' buffer sizes

Fig. 6.5 The adaptive network formation enables dynamic switching between U2B and U2U links along UAVs' trajectories. Nearby UAVs relay data for distant ones, reducing latency and energy use. Resource-aware coordination balances data transmission, prevents congestion, and minimizes mission time

metrics, though at the expense of slower convergence. Notably, Algorithm 1 surpasses dynamic-NF by achieving both superior reward and faster convergence, as evidenced in Fig. 6.6a. Figure 6.6b illustrates the UAVs' peak buffer occupancy dur-

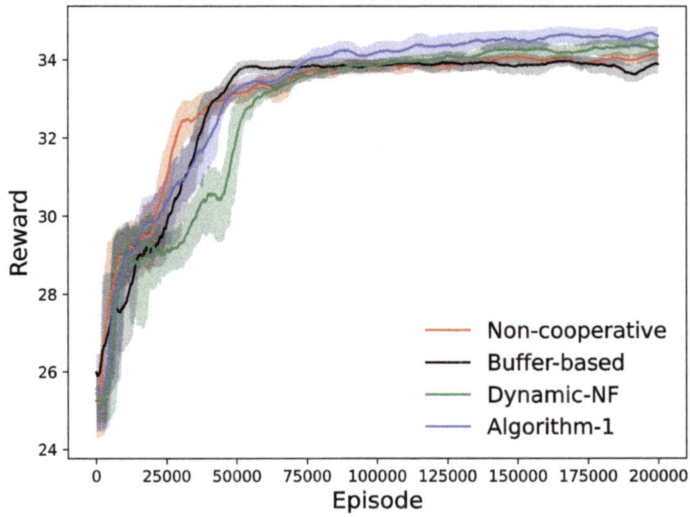

(a) Reward dynamics in learning

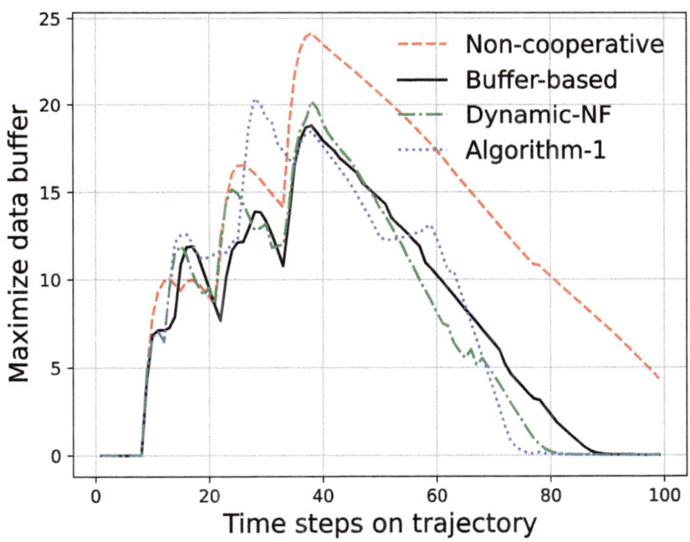

(b) UAVs' buffer sizes

Fig. 6.6 Adaptive network formation outperforms non-cooperative, buffer-based, and dynamic-NF baselines, achieving higher rewards and faster convergence. It enables faster data transmissions, reducing hovering time and energy use, and ensuring timely task completion

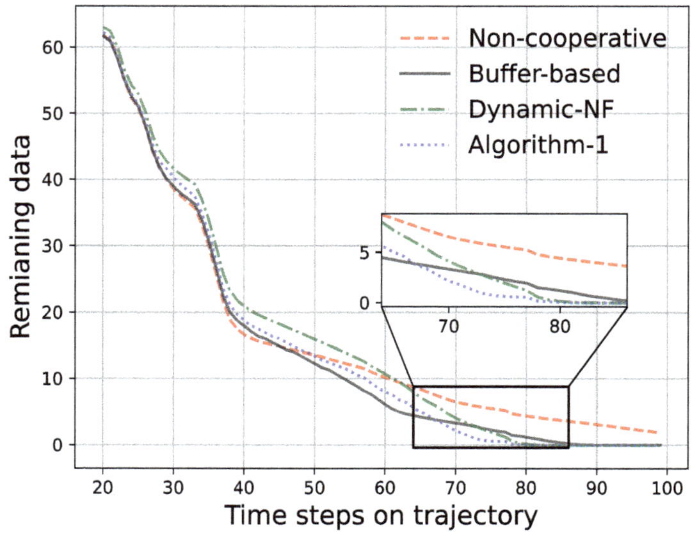

(a) Remaining data in the system

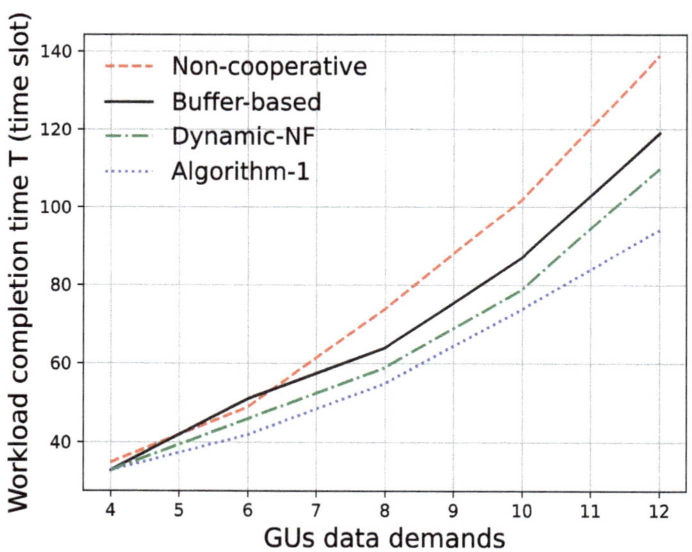

(b) Workload completion time

Fig. 6.7 While all baselines reduce buffer data, non-cooperative UAVs suffer from slow U2B rates. Cooperative methods use U2U links for faster offloading. Algorithm 1 clears data fastest, outperforming non-cooperative, buffer-based, and dynamic-NF methods, especially under high workloads. This highlights its effectiveness in resource-aware network reformation and workload balancing for timely task completion

ing flight. Adaptive network formation effectively keeps buffer levels low, whereas the non-cooperative strategy results in the highest peak, particularly for UAVs far from the BS with poor U2B channel quality. The buffer-based algorithm achieves the lowest maximum buffer size due to its exclusive focus on buffer management. Algorithm 1 maintains a comparable peak occupancy. More importantly, it enables all UAVs to complete data transmissions faster, indicating reduced hovering time and lower in-flight energy expenditure.

Figure 6.7a tracks the remaining data in both GUs' and UAVs' buffers over time. All methods show a decreasing trend. The non-cooperative strategy exhibits rapid initial reduction followed by a slow and long tail, indicating extended completion time due to U2B rate limitations. In contrast, the buffer-based and dynamic-NF methods leverage high-speed U2U links to enhance system throughput, significantly shortening task completion. Algorithm 1 achieves the fastest data clearance. As workload intensity increases, its advantage becomes more pronounced. This is further quantified in Fig. 6.7b, which plots task completion time against increasing GUs' workload demands. Algorithm 1 consistently achieves shorter completion times, with an increasing performance gap under heavier loads. This demonstrates its superior robustness in balancing UAVs' workloads and managing resource allocation through resource-aware network reformation during flight.

6.6 Discussions and Conclusions

This case study presents a hierarchical learning framework for joint optimization of UAVs' trajectories and network formation to enhance wireless sensing data collection and transmission. With trajectories initially fixed, network formation is dynamically adjusted based on real-time factors such as UAVs' energy levels and buffer occupancy. Whenever the network topology changes, UAVs' trajectories are refined using a multi-agent DRL approach. To accelerate learning, BO is integrated into the multi-agent DRL framework, using historical data to predict high-reward flying actions. This guidance reduces exploration of suboptimal maneuvers and significantly improves convergence speed. Simulations demonstrate that the proposed BO-MADDPG framework effectively coordinates mobility control and networking decisions, adapting the network structure in real time and achieving superior data collection performance compared to conventional methods.

References

1. L. Gupta, R. Jain, G. Vaszkun, Survey of important issues in UAV communication networks. IEEE Commun. Surv. Tut. **18**(2), 1123–1152 (2015)
2. S. Hayat, E. Yanmaz, R. Muzaffar, Survey on unmanned aerial vehicle networks for civil applications: a communications viewpoint. IEEE Commun. Surv. Tut. **18**(4), 2624–2661 (2016)

3. Y. Zhang, X. Lan, J. Ren, L. Cai, Efficient computing resource sharing for mobile edge-cloud computing networks. IEEE/ACM Trans. Netw. **28**(3), 1227–1240 (2020)
4. N. Zhao, W. Lu, M. Sheng, Y. Chen, J. Tang, F.R. Yu, K. Wong, UAV-assisted emergency networks in disasters. IEEE Wireless Commun. **26**(1), 45–51 (2019)
5. S. Gong, M. Wang, B. Gu, W. Zhang, D.T. Hoang, D. Niyato, Bayesian optimization enhanced deep reinforcement learning for trajectory planning and network formation in multi-UAV networks. IEEE Trans. Veh. Technol. **72**(8), 10933–10948 (2023)
6. M. Mozaffari, W. Saad, M. Bennis, M. Debbah, Mobile unmanned aerial vehicles (UAVs) for energy-efficient internet of things communications. IEEE Trans. Wireless Commun. **16**(11), 7574–7589 (2017)
7. M.A. Abd-Elmagid, H.S. Dhillon, Average peak age-of-information minimization in UAV-assisted IoT networks. IEEE Trans. Veh. Techn. **68**(2), 2003–2008 (2018)
8. S. Zhang, H. Zhang, B. Di, L. Song, Cellular UAV-to-X communications: design and optimization for multi-UAV networks. IEEE Trans. Wireless Commun. **18**(2), 1346–1359 (2019)
9. X. He, R. Jin, H. Dai, Multi-hop task offloading with on-the-fly computation for multi-UAV remote edge computing. IEEE Trans. Commun. **70**(2), 1332–1344 (2021)
10. S. Zhang, J. Yang, H. Zhang, L. Song, Dual trajectory optimization for a cooperative internet of UAVs. IEEE Commun. Lett. **23**(6), 1093–1096 (2019)
11. N.C. Luong, D.T. Hoang, S. Gong, D. Niyato, P. Wang, Y.-C. Liang, D.I. Kim, Applications of deep reinforcement learning in communications and networking: a survey. IEEE Commun. Surv. Tut. **21**(4), 3133–3174 (2019)
12. P. Luong, F. Gagnon, L.-N. Tran, F. Labeau, Deep reinforcement learning based resource allocation in cooperative UAV-assisted wireless networks. IEEE Trans. Wireless Commun. **20**(11), 7610–7625 (2021)
13. A. Feriani, E. Hossain, Single and multi-agent deep reinforcement learning for AI-enabled wireless networks: a tutorial. IEEE Commun. Surv. Tut. **23**(2), 1226–1252 (2021)
14. M. Alzenad, A. El-Keyi, H. Yanikomeroglu, 3-D placement of an unmanned aerial vehicle base station for maximum coverage of users with different QoS requirements. IEEE Wireless Commun. Lett. **7**(1), 38–41 (2017)
15. H. Luan, Y. Xu, D. Liu, Z. Du, H. Qian, X. Liu, X. Tong, Energy efficient task cooperation for multi-UAV networks: a coalition formation game approach. IEEE Access **8**, 149372–149384 (2020)
16. E.W. Frew, T.X. Brown, Airborne communication networks for small unmanned aircraft systems, *Proceedings of the IEEE*, vol. 96, no. 12 (2008)
17. U. Challita, W. Saad, Network formation in the sky: unmanned aerial vehicles for multi-hop wireless backhauling, in *Proceedings of IEEE GLOBECOM* (2017), pp. 1–6
18. Q. Chen, Joint position and resource optimization for multi-UAV-aided relaying systems. IEEE Access **8**, 10403–10415 (2020)
19. B. Liu, W. Zhang, W. Chen, H. Huang, S. Guo, Online computation offloading and traffic routing for UAV swarms in edge-cloud computing. IEEE Trans. Veh. Techn. **69**(8), 8777–8791 (2020)
20. F. Xiong, H. Zheng, L. Ruan, H. Wang, L. Tang, X. Dong, A. Li, Energy-saving data aggregation for multi-UAV system. IEEE Trans. Veh. Technol. **69**(8), 9002–9016 (2020)
21. O. Ghdiri, W. Jaafar, S. Alfattani, J.B. Abderrazak, H. Yanikomeroglu, Offline and online UAV-enabled data collection in time-constrained IoT networks. IEEE Trans. Green Commun. Netw. **5**(4), 1918–1933 (2021)
22. O. Bouhamed, H. Ghazzai, H. Besbes, Y. Massoud, A UAV-assisted data collection for wireless sensor networks: autonomous navigation and scheduling. IEEE Access **8**, 110446–110460 (2020)
23. L. Liu, K. Xiong, J. Cao, Y. Lu, P. Fan, K.B. Letaief, Average AoI minimization in UAV-assisted data collection with RF wireless power transfer: a deep reinforcement learning scheme. IEEE Internet Things J. **9**(7), 5216–5228 (2022)
24. M. Sun, X. Xu, X. Qin, P. Zhang, AoI-energy-aware UAV-assisted data collection for IoT networks: a deep reinforcement learning method. IEEE Internet Things J. **8**(24), 17275–17289 (2021)

25. J. Cui, Y. Liu, A. Nallanathan, Multi-agent reinforcement learning-based resource allocation for UAV networks. IEEE Trans. Wireless Commun. **19**(2), 729–743 (2019)
26. Q. Wang, W. Zhang, Y. Liu, Y. Liu, Multi-UAV dynamic wireless networking with deep reinforcement learning. IEEE Commun. Lett. **23**(12), 2243–2246 (2019)
27. H. Qie, D. Shi, T. Shen, X. Xu, Y. Li, L. Wang, Joint optimization of multi-UAV target assignment and path planning based on multi-agent reinforcement learning. IEEE Access **7**, 146264–146272 (2019)
28. Y. Zhang, Z. Mou, F. Gao, J. Jiang, R. Ding, Z. Han, UAV-enabled secure communications by multi-agent deep reinforcement learning. IEEE Trans. Veh. Technol. **69**(10), 11599–11611 (2020)
29. Z. Zhu, S. Wan, P. Fan, K.B. Letaief, Federated multiagent actor-critic learning for age sensitive mobile-edge computing. IEEE Internet Things J. **9**(2), 1053–1067 (2022)
30. Y. Zhang, Z. Mou, F. Gao, L. Xing, J. Jiang, Z. Han, Hierarchical deep reinforcement learning for backscattering data collection with multiple UAVs. IEEE Internet Things J. **8**(5), 3786–3800 (2020)
31. J. Hu, H. Zhang, L. Song, R. Schober, H.V. Poor, Cooperative internet of UAVs: distributed trajectory design by multi-agent deep reinforcement learning. IEEE Trans. Commun. **68**(11), 6807–6821 (2020)
32. F. Wu, H. Zhang, J. Wu, Z. Han, H.V. Poor, L. Song, Uav-to-device underlay communications: age of information minimization by multi-agent deep reinforcement learning. IEEE Trans. Commun. **69**(7), 4461–4475 (2021)
33. Y. Zeng, R. Zhang, Energy-efficient UAV communication with trajectory optimization. IEEE Trans. Wireless Commun. **16**(6), 3747–3760 (2017)
34. R. Lowe, Y. Wu, A. Tamar, J. Harb, P. Abbeel, I. Mordatch, Multi-agent actor-critic for mixed cooperative-competitive environments, in *Proceedings of International Conference on Neural Information Processing Systems (NeurIPS)* (2017), pp. 6382–6393
35. E. Brochu, V. M. Cora, N. de Freitas, A tutorial on Bayesian optimization of expensive cost functions, with application to active user modeling and hierarchical reinforcement learning. CoRR, abs/1012.2599 (2010). [Online]. Available: http://arxiv.org/abs/1012.2599
36. C. Rasmussen, C. Williams, *Gaussian Processes for Machine Learning*, Ser. Adaptive Computation and Machine Learning (MIT Press, Cambridge, MA, USA, 2006)
37. Z. Zhuang, J. Wang, Q. Qi, J. Liao, Z. Han, Adaptive and robust routing with Lyapunov-based deep RL in MEC networks enabled by blockchains. IEEE Internet Things J. **8**(4), 2208–2225 (2021)
38. M. Wang, Y. Long, S. Gong, J. Xu, Adaptive network formation and trajectory optimization for multi-UAV-assisted wireless data offloading, in *Proceedings of IEEE HPCC* (2021), pp. 961–967

Chapter 7
Lyapunov-Guided DRL for Stochastic AoI Minimization

7.1 Introduction

Stochastic resource optimization presents a significant challenge in dynamic wireless networks. Conventional optimization methods, which often rely on simplified models and exhibit limited adaptability, struggle to effectively manage uncertainty and satisfy long-term performance constraints. While the model-free DRL offers greater robustness and flexibility, it also face difficulties in guaranteeing system stability or strictly enforcing the stochastic constraints. To overcome these limitations, Lyapunov optimization can be integrated into the DRL framework, providing a mechanism to ensure system stability and control long-term performance metrics. Such an integration effectively achieves a balance between immediate rewards and long-term requirements, resulting in a more reliable and robust optimization process in stochastic environments. Our case studies in previous chapters have demonstrated that the model-based optimization methods can improve learning efficiency and reward performance of model-free DRL. The case study in this chapter further reveals that model-free DRL can be also embedded within an optimization framework and beneficial for stochastic resource allocation in wireless networks. The detailed analysis and algorithm design in the following sections were originally presented in our earlier work [1]. This chapter refines and reorganizes that content to ensure consistency, clarity, and coherence within the broader scope of the book.

7.1.1 Motivations and Contributions

The future IoT is evolving into an emerging technology for real-time and ubiquitous data sensing, enabling a wide range of applications including warehouse logistics, autonomous driving, and metaverse [2, 3]. In these latency-sensitive scenarios, the freshness of sensory data is critical to ensure system performance and reliability.

S. Gong et al., *Optimization-Driven Deep Reinforcement Learning for Wireless Networks*, Wireless Networks, https://doi.org/10.1007/978-3-032-22997-7_7

Timely delivery of sensing information from IoT devices to their intended receivers is essential to preserve its importance and freshness. AoI has been widely adopted as a key metric to quantitatively measure information freshness. Specifically, it measures the time elapsed since the most recent successful data update was generated at the sensing device [4]. A lower AoI value indicates more recent information, leading to enhanced situational awareness and more accurate decision-making. In contrast, a higher AoI value implies stale sensory data, which increases the risk of discrepancy between the observed system state and the actual system state. This obsolescence can severely degrade the performance of real-time monitoring and control systems, potentially compromising operational safety and efficiency [5, 6]. For instance, in autonomous driving, the sensor's delayed status updates may lead to misjudgment of traffic conditions and hazardous vehicle behavior. Similarly, in industrial IoT systems, stale environmental data leads to suboptimal automation and control decisions, directly impacting productivity and profitability. Given the inherent stochasticity of wireless channels, dynamic behavior of wireless devices, and unpredictable transmission delays, it remains a highly challenging problem to minimize AoI in practical wireless networks. Physical constraints such as limited energy resources and fluctuating network conditions further complicate the design of effective resource allocation strategies. These challenges motivate the development of intelligent and efficient resource allocation that can maintains low time-averaged AoI while satisfying stochastic resource constraints in dynamic IoT environments.

In this case study, we investigate the tailored design and application of a Lyapunov-guided DRL framework for a UAV-assisted wireless network, comprising one BS, multiple UAVs, and a large number of GUs. Both GUs and UAVs are equipped with semantic extraction modules, enabling information compression and efficient transmission of more important data. Our objective is to develop an efficient control strategy that jointly optimizes UAV-GU association, semantic extraction decisions, and UAVs' trajectories, aiming to minimize the time-averaged AoI across all GUs. Depending on real-time network conditions, each UAV first decides whether to perform semantic extraction on the raw sensing data received from GUs or to forward the data directly to the BS. Subsequently, the UAVs determine their next hovering locations based on current AoI levels and semantic control decisions. The joint optimization of UAV-GU association, trajectory planning, and semantic extraction policies constitutes a high-dimensional mixed-integer programming problem under stochastic constraints. To tackle this stochastic optimization, we first employ Lyapunov optimization to decompose the long-term problem into a sequence of tractable per-slot subproblems. Then, due to the computational complexity of these subproblems, we adopt a hybrid learning framework in each time slot. The PPO-based DRL method is applied to adapt UAV-GU association in the outer loop, while the inner-loop AO and SCA are used to optimize UAVs' trajectory planning and semantic extraction strategies, respectively.

Specifically, our main technical contributions in this case study are three fold. First, we propose a UAV-assisted semantic communication system capable of adaptively extracting semantic information at either the UAVs or the GUs, depending on real-time data traffic and channel conditions. For small-size raw data, GUs can directly transmit it to the BS via UAV-assisted relay communications. For large-size raw data, direct transmissions usually introduce substantial power consumption and time delays especially under poor channel conditions. In this case, GUs can extract salient semantic information before uplink data transmission. The UAVs not only act as the relay nodes to forward GUs' semantic information, but also serve as the edge servers to assist GUs' semantic extraction on demand. Second, we introduce semantic-aware AoI (SAoI) as a novel metric to jointly capture information freshness and semantic value at the BS. High semantic value helps preserve the integrity of recovered information, while low AoI supports timely and accurate decision-making. We formulate a long-term stochastic optimization problem aimed at minimizing the average SAoI through a joint optimization of UAV-GU association, semantic extraction, and UAVs' trajectory planning. Using a Lyapunov optimization framework, this stochastic problem is decomposed into a sequence of tractable subproblems in successive time slots. Third, we develop a Lyapunov-guided hierarchical PPO (Hily-PPO) framework to efficiently solve the SAoI minimization problem. The control subproblem in each time slot requires the optimization of UAV-GU association, semantic extraction, and UAVs' deployment strategies according to current AoI states. The model-free PPO algorithm is first applied to optimize the UAV-GU association strategy. Then, with fixed UAV-GU association, the model-based AO and SCA methods are employed to iteratively optimize semantic extraction and trajectory planning strategies. Extensive simulation results verify that the proposed Hily-PPO not only accelerates convergence but also achieves superior reward performance compared to typical baseline algorithms.

7.1.2 Related Works

7.1.2.1 UAV-Assisted AoI Minimization

UAV-assisted wireless networks have recently emerged as a promising technology for enabling real-time data communications across wide geographic areas [7, 8]. By leveraging LoS channels, UAVs can provide enhanced wireless service to GUs [9, 10]. However, UAVs' fast mobility also introduces dynamic and time-varying channel conditions, while competition for limited channel resources between UAVs and GUs further exacerbates transmission uncertainty. This leads to an increasing risks of data loss and degrading timeliness of information updates. This implies that AoI minimization in such networks is highly dependent on UAVs' trajectory planning and an efficient UAV-GU association strategy [11]. For instance, the authors in [12] investigated AoI-minimal trajectory planning by optimizing the UAV's hovering positions, proposing a path search algorithm that identifies efficient trajectories to maintain information freshness. Similarly, a double DQN algorithm was employed in [13] to effectively optimize the UAV's trajectory that enhances decision timeliness and reduce average AoI in latency-sensitive vehicular networks. The work in [14] focused on minimizing both AoI and energy consumption through coordinated multi-UAV trajectory planning, using a distributed DRL framework with inter-UAV message sharing. Further studies have extended this line of research under various constraints. The work in [15] proposed a DRL-based online scheduling and dynamic altitude control scheme to ensure information freshness at the BS, maintaining a low AoI even under poor channel conditions. Furthermore, focusing on IRS-assisted UAV networks, the authors in [16] proposed a joint optimization of GU's transmission scheduling, UAV's trajectory, and IRS's phase matrix to enhance AoI performance and transmission rates simultaneously.

7.1.2.2 UAV-Assisted Semantic Communication

Recent advances in semantic communication have shifted the focus of information delivery from traditional bit-level accuracy to context-aware semantic significance [17, 18]. In this framework, the transmitter extracts and encodes the essential meaning from raw data, referred to as semantic information, significantly reducing the volume of data that needs to be transmitted. When accurately reconstructed at the receiver, this semantic representation preserves an accurate interpretation of the original message, maintaining high information value. However, due to channel fluctuations or limited computational capabilities, imperfect reconstruction can result in semantic distortion at the receiver, degrading the fidelity and the practical value of the original information. This inherent challenge implies a critical trade-off study in such semantic communication systems. While more compact semantic representations can minimize transmission delay and bandwidth consumption, they often demand higher computational resources for accurate decoding and semantic recovery at the receiver. This creates the fundamental trade-off between transmission efficiency and reconstruction accuracy. Consequently, there is a growing need to jointly optimize timeliness quantified by AoI and semantic value of information. Balancing these two aspects is crucial for enhancing the performance of semantic-aware communication systems, particularly in delay-sensitive and resource-constrained applications such as UAV-assisted wireless networks.

Semantic communication enables UAVs to extract and interpret semantic features from GUs, significantly reducing data redundancy and enhancing the efficiency of information transmission. For instance, the authors in [19] employed a DRL algorithm to facilitate semantic transmissions of image and channel state information, achieving an effective trade-off between transmission latency and image classification accuracy. In [20], semantic information derived from raw visual perception was leveraged to encode environmental states, allowing UAVs to better understand and describe their surroundings and thereby navigate autonomously in complex and unknown network environments. Furthermore, the authors in [21] investigated semantic transmissions over dynamic wireless channels, proposing a game-theoretic multi-user resource allocation mechanism to enhance anti-interference capability and improve the overall quality of semantic communication services. Meanwhile, focusing on UAV-assisted semantic services for secondary users, the authors in [22] developed a hybrid DDQN-DDPG-based resource allocation algorithm that maximizes the spectrum efficiency of the secondary network by jointly optimizing subchannel assignment, semantic symbol allocation, UAV's transmit beamforming, and trajectory planning.

Semantic-aware information freshness aims to maintain not only up-to-date but also meaningful and contextually relevant information in communication systems. The authors in [23] introduced a knowledge graph-assisted semantic communication framework for time division multiple access system to enhance information freshness by extracting and transmitting high-level semantic representations instead of raw data. Meanwhile, the authors in [24] applied semantic communication to metaverse services to deliver immersive and realistic user experiences with reduced AoI. This method employs scene graph extraction and trading to capture environmental semantics, minimizing communication overhead while incentivizing efficient data updates. The work in [25] adopted AoI as a key metric to reflect semantic importance in status update systems, demonstrating that reliable semantic communication is achievable even under unreliable channel conditions. These works discussed above highlight the significant potential of integrating semantic awareness with information freshness metrics to enable more cost-effective and context-aware communication systems.

7.2 System Model and Problem Formulation

In this case study, we focus on a UAV-assisted semantic communication network with one BS, K GUs and M UAVs, as shown in Fig. 7.1. The sets of the GUs and the UAVs are denoted as $\mathcal{K} \triangleq \{1, 2, \ldots, K\}$ and $\mathcal{M} \triangleq \{1, 2, \ldots, M\}$, respectively. We assume that the GUs cannot be served by the BS directly due to the blockage of surrounding obstacles. The UAVs are deployed to collect sensing data from K GUs, randomly distributed in an open area, and then forward the collected sensing data to the BS for decision update. Each GU randomly generates varying-size raw data. For the small-size raw data, the UAVs can directly relay them to the BS without any processing. However, due to poor channel conditions or limited wireless resources, the transmission of large-size raw data results in high latency. In this case, the semantic extraction facilitates shorter delay by transmitting meaningful information. The GUs first extract the semantic information based on a local probability graph. If the GUs lack sufficient processing capacities, the UAVs can collect large-size raw data and then extract the semantic information based on edge probability graphs before transmitting it to the BS. Finally, the information recovery can be done at the BS due to its powerful processing capacity.

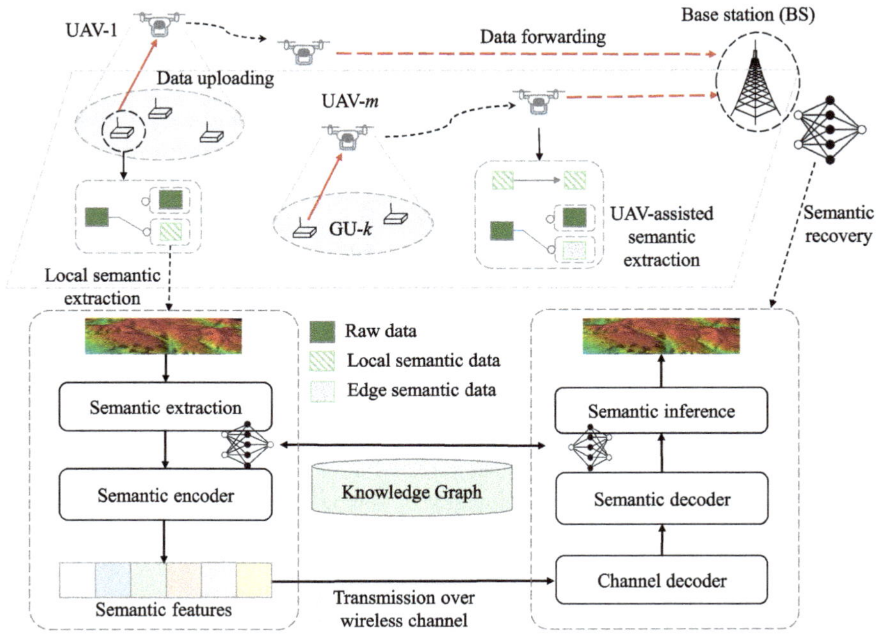

Fig. 7.1 A UAV-assisted semantic communication network where raw sensing data undergoes semantic extraction either locally at GUs or at UAVs before transmission, enabling efficient and low-latency communication through semantic information recovery at the BS

7.2.1 Extraction-Then-Transmission (EXT) Protocol

We devise a time-slotted extraction-then-transmission (EXT) protocol, depicted in Fig. 7.1. Each operational time frame is partitioned into I time slots with length $t_{\max}$. The set of all time slots is denoted by $\mathcal{I} \triangleq \{1, 2, \ldots, I\}$. All UAVs operate at a constant altitude. The two-dimensional location of UAV-m and GU-k in the i-th time slot are represented as $\boldsymbol{\ell}_m(i) = (x_m(i), y_m(i))$ and $\boldsymbol{q}_k = (x_k, y_k)$, respectively. Within a single time slot, the UAVs' positions are assumed to be fixed for modeling convenience. The BS's receiving antenna is located at $\boldsymbol{q}_0 = (x_0, y_0)$, and for notational consistency, it is treated as a special terrestrial node denoted as GU-0. Thus, the extended set of GUs including the BS is defined as $\tilde{\mathcal{K}} \triangleq \mathcal{K} \bigcup \{0\}$.

Each time slot under the EXT protocol can be further divided into a few sub-slots denoted as $\boldsymbol{t}_k(i) \triangleq (t^l_{k,e}(i), t_{k,s}(i), t^u_{k,e}(i), t_{k,f}(i), t_{k,r}(i))$, corresponding to five sequential phases for semantic extraction at the GUs, data upload from GUs to UAVs, edge-based semantic extraction at UAVs, data forwarding from UAVs to the BS, and semantic recovery at the BS. If $t^l_{k,e}(i) = 0$, GU-k skips local processing and transmits the raw data to UAVs. Otherwise, it performs semantic extraction during $t^l_{k,e}(i)$. Depending on this decision, GU-k uploads either raw data or processed semantic information to the associated UAV in the subsequent sub-slot $t_{k,s}(i)$. When

$t^u_{k,e}(i) = 0$, UAVs receive either pre-extracted semantics or raw data without further processing. Conversely, if $t^u_{k,e}(i) > 0$, UAVs act as edge computing nodes to help extract semantic content, thereby reducing GUs' computational load during $t^u_{k,e}(i)$. In the sub-slot $t_{k,f}(i)$, UAVs forward the received or processed information to the BS. Finally, during $t_{k,r}(i)$, the BS reconstructs the original message from the received semantic representation. The total duration for information processing is given by

$$\tilde{t}_k(i) = t^l_{k,e}(i) + t_{k,s}(i) + t^u_{k,e}(i) + t^l_{k,r}(i) + t^u_{k,r}(i) \leq t_{\max}. \tag{7.1}$$

For any scheduled GU-k, its entire transmission and processing must be completed within one time slot, i.e., $\tilde{t}_k(i) \leq t_{\max}$.

UAVs' trajectory planning must ensure safe operation by avoiding collisions and respecting the maximum speed $v_{\max}$ in each time slot $i \in \mathcal{I}$. These constraints are expressed as follows:

$$\|\boldsymbol{\ell}_m(i) - \boldsymbol{\ell}_{m'}(i)\| \geq d_{\min}, \quad \forall m, m' \in \mathcal{M} \text{ and } m \neq m', \tag{7.2a}$$

$$\|\boldsymbol{\ell}_m(i) - \boldsymbol{\ell}_m(i-1)\| \leq t_{\max} v_{\max}, \quad \forall m \in \mathcal{M}, \tag{7.2b}$$

where $d_{\min}$ defines the minimum safe separation between any two UAVs, and $v_{\max}$ is the maximum allowable flying speed [26]. For any $m \in \mathcal{M}$ and $k \in \tilde{\mathcal{K}}$, the complex channel vector $\mathbf{h}_{m,k}(i)$ between UAV-m and GU-k at slot i is modeled as follows:

$$h_{m,k}(i) = \frac{\sqrt{\xi}}{d_{s,m,k}(i)} \left(\sqrt{\frac{g_0}{g_0 + 1}} \bar{g}_{m,k}(i) + \sqrt{\frac{1}{g_0 + 1}} \tilde{g}_{m,k}(i) \right), \tag{7.3}$$

where ξ is the channel power gain at a reference distance. The Rician factor g_0 accounts for both the dominant LoS component $\bar{g}_{m,k}(i)$ and the NLoS fading component $\tilde{g}_{m,k}(i)$ [27].

A binary variable $\beta_{m,k}(i)$ is introduced to represent the association between UAV-m and GU-k in time slot i, where $\beta_{m,k}(i) = 1$ indicates an active link, and 0 otherwise. To prevent signal interference during the uplink phase, each UAV serves at most one GU and each GU connects to no more than one UAV in each time slot. This leads to the following scheduling constraints:

$$\sum_{m \in \mathcal{M}} \beta_{m,k}(i) \leq 1 \text{ and } \sum_{k \in \mathcal{K}} \beta_{m,k}(i) \leq 1, \tag{7.4a}$$

$$\beta_{m,k}(i) \in \{0, 1\}, \ \forall m \in \mathcal{M} \text{ and } k \in \mathcal{K}. \tag{7.4b}$$

7.2.2 Semantic Information Extraction

The goal of semantic extraction is to capture the essential meaning of the source data, enabling more efficient communication by eliminating redundant information. Instead of transmitting the full raw data, only essential semantic features are conveyed, which can be extracted either locally at the GUs or offloaded to UAVs for edge-assisted processing. This extraction is performed by a semantic encoder that leverages task-specific requirements and pre-shared knowledge bases [28]. Specifically, GU-k can derive semantic content $s_k^l(i)$ from its original raw data $D_k(i)$ through local computation. To quantify the extent of data reduction, we introduce the concept of semantic extraction depth, defined as $\rho_k^l(i) \triangleq s_k^l(i)/D_k(i)$, representing the ratio of extracted semantic information to the original data size at GU-k. The time required for GU-k to perform local semantic extraction is modeled as follows:

$$t_{k,e}^l(i) = \sum_{m\in\mathcal{M}} \frac{\beta_{m,k}(i)}{f_k^l(i)} y_k^{l,e}(D_k(i), \rho_k^l(i)), \tag{7.5}$$

where $f_k^l(i)$ denotes the computational resources (e.g., CPU cycle frequency) allocated to GU-k and $y_k^{l,e}(D_k(i), \rho_k^l(i))$ represents the total CPU cycles needed for the extraction process. This value can be evaluated as follows [28]:

$$y_k^{l,e}(D_k(i), \rho_k^l(i)) = y_k^{l,o}(D_k(i)) + B_{1k}\left(1 - \rho_k^l(i)\right)^{B_{2k}}, \tag{7.6}$$

where $y_k^{l,o}(D_k(i)) = B_{0k}D_k(i)$ accounts for the construction of shared knowledge from the raw data and the second term captures the computational effort dependent on the extraction depth $\rho_k^l(i)$. The parameters (B_{0k}, B_{1k}, B_{2k}) are determined empirically through curve fitting based on simulation data [28].

Similarly, when UAV-m performs semantic extraction, the resulting semantic information is denoted as $s_{m,k}^u(i)$ and the corresponding extraction depth is given by $\rho_{m,k}^u(i)$. The time consumed by UAV-m for processing the raw data $D_k(i)$ is given as follows:

$$t_{k,e}^u(i) = \sum_{m\in\mathcal{M}} \frac{\beta_{m,k}(i)}{f_{m,k}^u(i)} y_m^{u,e}(D_k(i), \rho_{m,k}^u(i)), \tag{7.7}$$

where $y_m^{u,e}(D_k(i), \rho_{m,k}^u(i))$ is the number of CPU cycles required to extract $s_{m,k}^u(i)$ and $f_{m,k}^u(i)$ is the computational capacity of UAV-m. The computation demand at the UAV-m can be approximated using a similar function:

$$y_m^{u,e}(D_k(i), \rho_{m,k}^u(i)) = y_m^{u,o}(D_k(i)) + C_{1m}(1 - \rho_{m,k}^u(i))^{C_{2m}}, \tag{7.8}$$

where $y_m^{u,o}(D_k(i)) = C_{0m}D_k(i)$ and the coefficients (C_{0m}, C_{1m}, C_{2m}) are experimentally derived constants. We assume that semantic extraction does not occur simultaneously at both the GU and the UAV for the same data. Specifically, UAVs only initiate semantic processing when they receive unprocessed raw data from a GU. This exclusivity is enforced by the following constraints:

$$\rho_k^l(i)\rho_{m,k}^u(i) = 0,\ \rho_k^l(i) \in [0, 1],\ \rho_{m,k}^u(i) \in [0, 1],\ \forall k \in \mathcal{K}, m \in \mathcal{M}. \tag{7.9}$$

7.2.3 *Information Transmission and Recovery*

Once semantic features are extracted, they are transmitted over the wireless medium using standard bit-based modulation and demodulation schemes. This ensures compatibility with existing communication systems while enabling efficient packaging and reliable delivery of semantic content. Given GU-k's transmit power p_k^o, the SNR observed at UAV-m is expressed as $\gamma_{m,k}(i) = p_k^o \left|h_{m,k}(i)\right|^2 / \sigma_m^2$, where σ_m^2 represents the noise power at the UAV. Consequently, the data rate from GU-k to UAV-m is $r_{m,k}^s(i) = \log_2(1 + \gamma_{m,k}(i))$, and the corresponding transmission time is as follows:

$$t_{k,s}(i) = \sum_{m\in\mathcal{M}} \frac{\beta_{m,k}(i)}{r_{m,k}^s(i)} \rho_k^l(i) D_k(i). \tag{7.10}$$

Likewise, with UAV-m transmitting at power p_m^u, the data rate from UAV-m to the BS is $r_{m,0}^f(i) = \log_2(1 + p_m^u \left|h_{m,0}^H(i)\right|^2 / \sigma_0^2)$, leading to the following forward transmission time:

$$t_{k,f}(i) = \sum_{m\in\mathcal{M}} \frac{\beta_{m,k}(i)}{r_{m,0}^f(i)} (\rho_k^l(i) + \rho_{m,k}^u(i)) D_k(i). \tag{7.11}$$

Note that the data forwarded by UAVs includes both the semantics extracted locally by the GUs and those processed at the UAVs themselves.

The received semantic signal at the BS may be corrupted by noise, potentially impairing accurate decoding. The BS first decodes the incoming bit stream and then employs semantic inference to reconstruct the original raw data $D_k(i)$, correcting any distortion introduced during wireless transmission. The availability of shared knowledge bases, common to both the transmitters (GUs or UAVs) and the receiver (BS), ensures consistency and accuracy in the reconstructed information. The BS's recovery process can be based on either the GUs' locally extracted semantic data $s_k^l(i)$ or the UAV-assisted semantics $s_{m,k}^u(i)$. The BS's computational demand for semantic recovery, in terms of CPU cycles, depends on the combined extraction depths $(\rho_k^l(i), \rho_{m,k}^u(i))$, which is modeled as follows:

$$y_k^r(\rho_k^l(i), \rho_{m,k}^u(i)) = B_{3k} \left(\rho_k^l(i) + \rho_{m,k}^u(i)\right)^{-B_{4k}}, \tag{7.12}$$

where (B_{3k}, B_{4k}) can be experimentally determined [28]. Accordingly, the time required to recover the data from GU-k is given by

$$t_{k,r}(i) = \sum_{m \in \mathcal{M}} \frac{\beta_{m,k}(i)}{c_k(i)} y_k^r(\rho_k^l(i), \rho_{m,k}^u(i)), \tag{7.13}$$

with $c_k(i)$ denoting the computation resources allocated by the BS for processing GU-k's data. Generally, a higher value of $\rho_k^l(i)$ or $\rho_{m,k}^u(i)$ implies that a larger portion of the original raw data is preserved during extraction, which supports more accurate reconstruction at the receiver. When $\rho_k^l(i) = 1$ or $\rho_{m,k}^u(i) = 1$, no semantic compression is applied, reducing the system to conventional bit-level transmission.

7.3 Long-Term Minimization of Semantic-Aware AoI

7.3.1 AoI Dynamics

AoI measures the elapsed time since the most recently received data packet was successfully reconstructed at the destination. As illustrated in Fig. 7.2, if GU-k is not connected to any UAV during time slot i, i.e., $\beta_{m,k}(i) = 0$ for all m, it cannot transmit its data to the BS through a UAV relay. Consequently, the AoI for GU-k increases

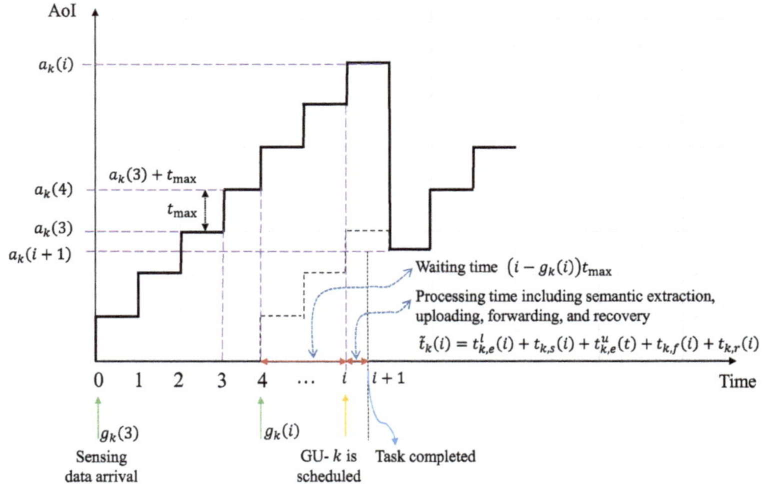

Fig. 7.2 GU-k's AoI evolves dynamically over time. AoI measures data freshness by tracking the time since the last successful packet delivery. If GU-k is not scheduled for transmission, its AoI increases by one slot. When it is scheduled, AoI is reset to the sum of the packet's waiting time and total processing delay, including semantic extraction, uplink transmission, forwarding, and recovery at the receiver

by one time slot, i.e., $a_k(i+1) = a_k(i) + t_{\max}$. Conversely, when $\beta_{m,k}(i) = 1$, UAV-m collects the data from GU-k and forwards it to the BS, allowing the AoI to be refreshed. In this case, the updated AoI at the next time slot is determined by the total latency experienced by the transmitted packet, which includes both its waiting time and end-to-end processing delay, as captured in the second term of (7.14). The waiting time reflects the duration between the generation of the most recent data packet at GU-k and the start of its transmission. Here, $\tilde{g}_k(i)$ in (7.14) denotes the time index when the data packet transmitted by GU-k in slot i was originally generated. Thus, the waiting period amounts to $(i - \tilde{g}_k(i))t_{\max}$. The processing time $\tilde{t}_k(i)$ encompasses all stages from data extraction to final recovery at the BS, including local semantic extraction at GU-k, uplink transmission from GU-k to UAV-m, potential edge-based semantic extraction at the UAV, forwarding from UAV-m to the BS, and semantic reconstruction at the BS. Therefore, upon successful transmission, the AoI is reset to $a_k(i+1) = (i - \tilde{g}_k(i))t_{\max} + \tilde{t}_k(i)$. For brevity, the AoI evolution for each GU can be expressed in a compact form as follows:

$$a_k(i+1) = \left(1 - \sum_{m \in \mathcal{M}} \beta_{m,k}(i)\right)(a_k(i) + t_{\max}) \\ + \sum_{m \in \mathcal{M}} \beta_{m,k}(i)\left(\left(i - \tilde{g}_k(i)\right)t_{\max} + \tilde{t}_k(i)\right), \quad \forall k \in \mathcal{K}. \tag{7.14}$$

Obsolete information may compromise system performance and lead to inaccurate decisions. To maintain data freshness, we impose an average AoI constraint, denoted as $a_{\max}$, such that:

$$\lim_{I \to \infty} \frac{1}{I} \sum_{i=0}^{I-1} \mathbb{E}\left[a_k(i+1)\right] \leq a_{\max}, \quad \forall k \in \mathcal{K}. \tag{7.15}$$

The expectation accounts for stochastic variations in channel conditions and the probabilistic nature of UAV-GU association decisions.

7.3.2 *Value of Semantic Information*

Semantic extraction enables GUs and UAVs to minimize redundant data during transmission, thereby improving communication efficiency and reducing latency or AoI. However, aggressive compression also makes the transmitted semantic information more susceptible to noise corruption over wireless channels. This can lead to challenges in accurately reconstructing the original content at the BS, potentially increasing the computational burden for recovery or degrading the fidelity of the reconstructed data. As a result, the overall usefulness of the received semantic information may be diminished despite its faster delivery. To capture this trade-off, we adopt a utility-based model for the value of semantic information, inspired by [29]. The semantic value for GU-k in time slot i is defined as:

$$v_k(i) = \sum_{m \in \mathcal{M}} \beta_{m,k}(i) \left(1 - e^{-B_{5k} D_k(i) \left(\rho_k^l(i) + \rho_{m,k}^u(i)\right)}\right), \tag{7.16}$$

where $B_{5k} \geq 0$ is a constant parameter characterizing the sensitivity of value to extraction depth. A higher $\rho_k^l(i)$ or $\rho_{m,k}^u(i)$ implies less compression, preserving more data content but increasing transmission and processing latency, which negatively impacts information freshness. Conversely, lower extraction depths result in greater compression, which improves AoI but makes semantic recovery more challenging, thereby reducing the reliability and value of information at the BS. This highlights the critical need to balance between minimizing AoI for freshness and maintaining sufficient semantic content to ensure high recovery accuracy and information value.

7.3.3 SAoI Minimization

Our objective is to minimize the overall semantic-aware age of information (SAoI) across all GUs by simultaneously optimizing three key decision variables: the UAV-GU association strategy $\boldsymbol{\beta} \triangleq \{\beta_{m,k}(i)\}_{k \in \mathcal{K}, m \in \mathcal{M}, i \in \mathcal{I}}$, the semantic extraction depths $\boldsymbol{\rho} \triangleq \{\rho_k^l(i), \rho_{m,k}^u(i)\}_{k \in \mathcal{K}, m \in \mathcal{M}, i \in \mathcal{I}}$, and the flight trajectories of the UAVs $\boldsymbol{\ell} \triangleq \{\ell_m(i)\}_{m \in \mathcal{M}, i \in \mathcal{I}}$. To account for the inherent trade-off between information freshness and the utility of semantic content, we define the averaged SAoI as follows:

$$\bar{A}(\boldsymbol{\beta}, \boldsymbol{\rho}, \boldsymbol{\ell}) = \lim_{I \to \infty} \frac{1}{IK} \mathbb{E}\left[\sum_{i \in \mathcal{I}} \sum_{k \in \mathcal{K}} \omega_1 a_k(i+1) - \omega_2 v_k(i)\right]. \tag{7.17}$$

This metric integrates both the timeliness of information, reflected in the AoI $a_k(i+1)$, and the quality of semantics, captured by the value function $v_k(i)$, with weighting factors ω_1 and ω_2 balancing their contributions. The expectation accounts for the stochastic nature of system dynamics influenced by the joint control policies. Thus, the SAoI minimization is formulated as:

$$\min_{\boldsymbol{\beta}, \boldsymbol{\rho}, \boldsymbol{\ell}} \bar{A}(\boldsymbol{\beta}, \boldsymbol{\rho}, \boldsymbol{\ell}), \quad s.t. \ (7.1)-(7.15). \tag{7.18}$$

Problem (7.18) is non-trivial due to the mixed discrete-continuous nature of the decision space and the spatio-temporal dependencies introduced by UAVs' mobility. The coupling among information freshness, semantic value, and the optimization variables introduces significant complexity. Specifically, the UAV-GU association $\boldsymbol{\beta}$ involves discrete combinatorial decisions, while the trajectory design $\boldsymbol{\ell}$ and semantic extraction control $\boldsymbol{\rho}$ are continuous variables that are interdependent across both space and time. Even when the UAV-GU association policy is fixed, the remaining subproblems remain highly coupled due to the dynamic and spatially varying channel conditions.

7.3.4 Lyapunov Decomposition

To tackle the SAoI minimization problem (7.18), we employ the Lyapunov optimization framework [30] to decompose the long-term stochastic problem into a sequence of per-slot optimization tasks. A key step involves transforming the average AoI constraint in (7.15) into an equivalent stability condition for a virtual queue, as stated in the following proposition.

Proposition 7.1 *For each GU-k, we introduce a virtual queue $Q_k(i)$ initialized to zero, i.e., $Q_k(0) = 0$, with its evolution governed by:*

$$Q_k(i+1) = \max\left[Q_k(i) - a_{\max}, 0\right] + a_k(i+1). \tag{7.19}$$

If the queue $Q_k(i)$ is mean rate stable, meaning $\lim_{i\to\infty} \frac{\mathbb{E}[|Q_k(i)|]}{i} = 0$, then the original constraint (7.15) is guaranteed to be satisfied.

The proof of Proposition 7.1 follows standard arguments in Lyapunov optimization and is detailed in [11]. Let $\mathbf{Q}(i) = (Q_1(i), \ldots, Q_K(i))$ denote the vector of all virtual AoI queue states. We define a quadratic Lyapunov function to quantify the system state and assess constraint satisfaction:

$$L\big(\mathbf{Q}(i)\big) = \frac{1}{2}\sum_{k\in\mathcal{K}} \big(Q_k(i)\big)^2, \quad \forall i \in \mathcal{I}. \tag{7.20}$$

This function is non-negative and increases with the magnitude of the virtual queues. The factor $1/2$ is included for analytical convenience in subsequent derivations. A smaller value of $L\big(\mathbf{Q}(i)\big)$ indicates that the virtual queues are well-controlled, implying better adherence to the average AoI requirement in (7.15). Conversely, large queue backlogs lead to a higher Lyapunov function value, signaling potential instability.

Next, we consider the Lyapunov drift, which captures the expected change in the Lyapunov function over one time slot:

$$\Delta_L\big(\mathbf{Q}(i)\big) = \mathbb{E}\big[L\big(\mathbf{Q}(i+1)\big) - L\big(\mathbf{Q}(i)\big)\big|\mathbf{Q}(i)\big]. \tag{7.21}$$

Minimizing this drift helps stabilize the virtual queues. We also require to reduce the SAoI of all GUs to maintain high information freshness and value. To balance these objectives, we define a drift-plus-penalty function as follows:

$$T(\mathbf{Q}(i)) \triangleq \Delta_L(\mathbf{Q}(i)) + V \sum_{k\in\mathcal{K}} \mathbb{E}\big[\omega_1 a_k(i+1) - \omega_2 v_k(i) \big| \mathbf{Q}(i)\big], \tag{7.22}$$

where $V \geq 0$ is a control weight that regulates the trade-off between queue stability and semantic information quality. By minimizing $T(\mathbf{Q}(i))$ at each time slot, we effectively address both the long-term constraint and the immediate performance objective. The following proposition provides an upper bound on the drift-plus-penalty $T(\mathbf{Q}(i))$, enabling a tractable per-slot optimization.

Proposition 7.2 *Given the virtual queue states $\mathbf{Q}(i)$, the function $T(\mathbf{Q}(i))$ is bounded above as:*

$$T(\mathbf{Q}(i)) \leq B + U(\boldsymbol{\beta}(i), \boldsymbol{\rho}(i), \boldsymbol{\ell}(i)), \tag{7.23}$$

where B is a finite constant defined by

$$B = \sum_{k\in\mathcal{K}} \left[\frac{1}{2}\left(a_{\max}^2 + (a_k(i) + t_{\max})^2\right) - Q_k(i) a_{\max}\right],$$

and $U(\boldsymbol{\beta}(i), \boldsymbol{\rho}(i), \boldsymbol{\ell}(i))$ is given by

$$U(\boldsymbol{\beta}(i), \boldsymbol{\rho}(i), \boldsymbol{\ell}(i)) = \sum_{k\in\mathcal{K}} \mathbb{E}\Big[(Q_k(i) + V\omega_1) a_k(i+1) - V\omega_2 v_k(i) | \mathbf{Q}(i)\Big],$$

with $a_k(i+1)$ updated according to (7.14).

The derivation of Proposition 7.2 is analogous to [11]. Since $Q_k(i)$ and $a_k(i)$ are known at the start of time slot i, the term B is constant and can be ignored in optimization. Therefore, minimizing the upper bound in (7.23) is equivalent to minimizing $U(\boldsymbol{\beta}(i), \boldsymbol{\rho}(i), \boldsymbol{\ell}(i))$. Dropping the time index for notational simplicity and omitting constant terms, we arrive at the following per-slot optimization problem:

$$\begin{aligned}
\min_{\boldsymbol{\beta},\boldsymbol{\rho},\boldsymbol{\ell}} \quad & \sum_{k\in\mathcal{K}} (Q_k + V\omega_1)\Bigg(\Big(1 - \sum_{m\in\mathcal{M}} \beta_{m,k}\Big)(a_k + t_{\max}) \\
& + \sum_{m\in\mathcal{M}} \beta_{m,k}\Big((i - \tilde{g}_k) t_{\max} + \tilde{t}_k\Big)\Bigg) - V\omega_2 v_k \\
\text{s.t.} \quad & (7.1)\text{–}(7.13),
\end{aligned} \tag{7.24}$$

which remains a challenging mixed-integer problem due to the discrete UAV-GU association variables and the coupled continuous decisions on trajectories and semantic extraction depths.

7.4 Lyapunov-Guided Hierarchical DRL Framework

The Lyapunov-guided hierarchical PPO framework (denoted as Hily-PPO) is shown in Fig. 7.3. For each time slot, we address the per-slot optimization problem (7.24) through a hierarchical PPO framework, as illustrated in Fig. 7.3. This framework integrates model-free PPO for UAV-GU association $\{\boldsymbol{\beta}(i)\}$ and model-based optimization for joint trajectory planning $\{\boldsymbol{\ell}(i)\}$ and semantic extraction control $\{\boldsymbol{\rho}(i)\}$. In every iteration, the UAVs first decide their associated GUs based on historical observations of AoI evolution. With the association strategy fixed, the continuous variables, i.e., UAVs' trajectories and semantic extraction depths, are optimized efficiently using the SCA method. After computing the full action set $(\boldsymbol{\beta}, \boldsymbol{\rho}, \boldsymbol{\ell})$, the BS executes the policy and observes the updated system state. Following this decision cycle, the virtual AoI queues are updated in preparation for the next time slot.

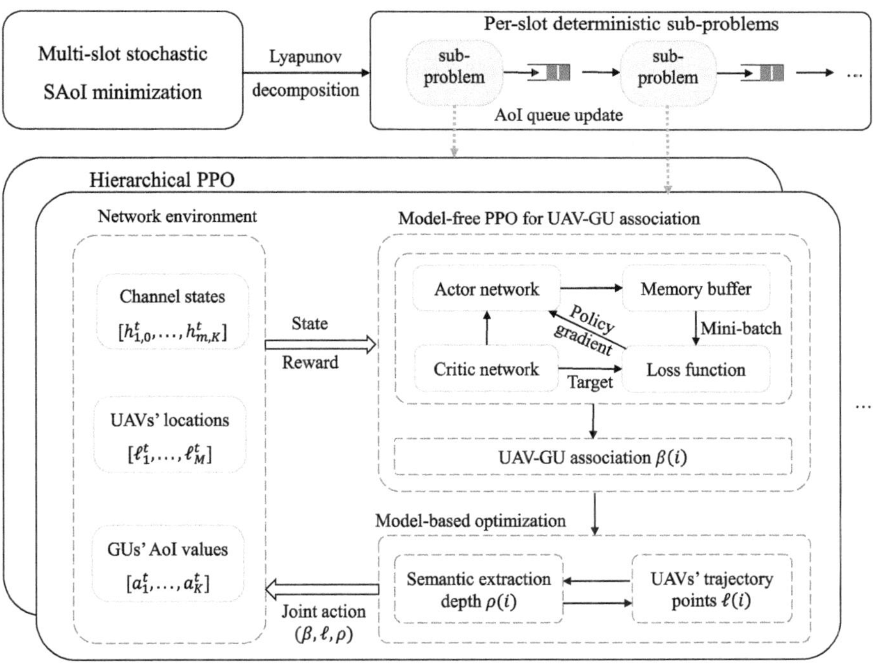

Fig. 7.3 The proposed Hily-PPO algorithm framework, relying on Lyapunov optimization to decompose multi-slot stochastic SAoI minimization problem into a series of per-slot deterministic sub-problems. For each sub-problem, model-free PPO is used to adapt UAV-GU association strategy $\boldsymbol{\beta}$ and then model-based AO method is used to optimize UAVs' trajectories $\boldsymbol{\ell}$ and semantic extraction depths $\boldsymbol{\rho}$

7.4.1 Model-Free PPO for UAV-GU Association

The BS determines the UAV-GU association by considering the global system state, including UAVs' positions, channel conditions, and the current AoI levels of all GUs. Once determined, this UAV-GU association strategy is distributed to individual UAVs for execution. We formulate this control task as a MDP, represented by the tuple $(\mathcal{S}, \mathcal{A}, \mathcal{R})$. The state vector $\boldsymbol{s}^t \in \mathcal{S}$ at decision epoch t includes the channel gains between all UAVs and GUs $\boldsymbol{h}^t \triangleq [h^t_{1,0}, \ldots, h^t_{m,K}]$, the spatial coordinates of UAVs $\boldsymbol{\ell}^t \triangleq [\ell^t_1, \ldots, \ell^t_M]$, and the instantaneous AoI values of all GUs $\boldsymbol{a}^t \triangleq [a^t_1, \ldots, a^t_K]$. Thus, the state is defined as $\boldsymbol{s}^t = \{\boldsymbol{h}^t, \boldsymbol{\ell}^t, \boldsymbol{a}^t\}$. At each step, the action $\boldsymbol{x}^t \in \mathcal{A}$ corresponds to selecting a UAV-GU association matrix $\boldsymbol{x}^t \triangleq [\beta^t_{1,1}, \ldots, \beta^t_{M,K}]$, which determines connectivity between UAVs and GUs under the current state $\boldsymbol{s}^t$. Reward function $(\boldsymbol{s}^t, \boldsymbol{x}^t) \in \mathcal{R}$ should be aligned with the objective of minimizing SAoI. Thus, the immediate reward is designed as $r^t(\boldsymbol{s}^t, \boldsymbol{x}^t) = U(\boldsymbol{\beta}^t, \boldsymbol{\rho}^t, \boldsymbol{\ell}^t)$, reflecting the utility derived from the selected action in terms of AoI reduction and semantic value preservation.

PPO offers a robust approach for policy learning within the actor-critic architecture by enforcing stable updates through a constraint on the KL divergence between successive policies [31]. Specifically, the old policy $\pi_{\theta'}$ interacts with the environment to collect experience tuples. These samples are then used to update the new policy π_θ while ensuring that the change remains within acceptable bounds. The update process is governed by the following constrained optimization problem:

$$\max_{\theta} \; E_{\pi_{\theta'}}\left[\frac{\pi_\theta(\boldsymbol{x}^t|\boldsymbol{s}^t)}{\pi_{\theta'}(\boldsymbol{x}^t|\boldsymbol{s}^t)} A_{\pi_{\theta'}}(\boldsymbol{s}^t, \boldsymbol{x}^t)\right] \tag{7.25}$$

$$s.t. \; E_{\pi_{\theta'}}[D_{KL}(\pi_\theta, \pi_{\theta'})] \leq \delta_{KL}, \tag{7.26}$$

where δ_{KL} sets a threshold on the expected KL divergence to prevent overly aggressive policy changes. The advantage function $A_{\pi_{\theta'}}(\boldsymbol{s}^t, \boldsymbol{x}^t)$ measures the relative benefit of taking action $\boldsymbol{x}^t$ in state $\boldsymbol{s}^t$ compared to the average outcome under the current policy. To simplify implementation, PPO replaces the hard constraint with a clipped surrogate objective, leading to the reformulated objective $\max_\theta E_{\pi_{\theta'}}\left[\tilde{J}(\pi_\theta) - \gamma_{KL} D_{KL}(\pi_\theta, \pi_{\theta'})\right]$, where $\tilde{J}(\pi_\theta)$ integrates the clipping mechanism to stabilize training:

$$\tilde{J}(\pi_\theta) = E_{\pi_{\theta'}}\left[\min\left(\rho_\theta A_{\pi_{\theta'}}, \text{clip}(\rho_\theta, 1-\epsilon, 1+\epsilon) A_{\pi_{\theta'}}\right)\right], \tag{7.27}$$

with $\rho_\theta = \frac{\pi_\theta(\boldsymbol{x}^t|\boldsymbol{s}^t)}{\pi_{\theta'}(\boldsymbol{x}^t|\boldsymbol{s}^t)}$ and $\epsilon > 0$ defining the clipping range. Denoting $\hat{A}_{\pi_{\theta'}} = \text{clip}(\rho_\theta, 1-\epsilon, 1+\epsilon) A_{\pi_{\theta'}}$, the effective advantage is adjusted as follows:

$$\hat{A}_{\pi_{\theta'}} = \begin{cases} (1+\epsilon) A_{\pi_{\theta'}}, & \text{if } \rho_\theta \geq 1+\epsilon \text{ and } A_{\pi_{\theta'}} > 0, \\ (1-\epsilon) A_{\pi_{\theta'}}, & \text{if } \rho_\theta \leq 1-\epsilon \text{ and } A_{\pi_{\theta'}} < 0, \\ \rho_\theta A_{\pi_{\theta'}}, & \text{otherwise.} \end{cases} \tag{7.28}$$

This mechanism ensures that when an action yields positive advantage, the policy update increases its likelihood, but only up to a multiplicative factor of $1+\epsilon$. Conversely, for actions with negative advantage, their selection probability is reduced, but not beyond a factor of $1-\epsilon$, thus maintaining sufficient exploration and preventing catastrophic policy degradation.

7.4.2 Optimizing Trajectory and Semantic Extraction

With the UAV-GU association $\boldsymbol{\beta} \triangleq \{\beta_{m,k}\}_{k\in\mathcal{K},m\in\mathcal{M}}$ fixed, the original problem (7.24) is decoupled into two subproblems for semantic extraction control and UAVs' positioning by using the AO method. First, assuming the UAVs' locations are known, we formulate the semantic extraction subproblem. The presence of the bilinear equality constraint $\rho_k^l(i)\rho_{m,k}^u(i)=0$ introduces non-convexity and complicates direct optimization. To address this, we relax the constraint by incorporating a penalty term into the objective. This leads to the following reformulated and tractable version:

$$\begin{aligned}\min_{\rho} &\sum_{k\in\mathcal{K}} \left[(Q_k+V\omega_1)\Gamma_1(\rho_k^l,\rho_{m,k}^u)-V\omega_2\Gamma_2(\rho_k^l,\rho_{m,k}^u)\right]\\ &+\omega_0\sum_{m\in\mathcal{M}}\sum_{k\in\mathcal{K}}\varrho_{m,k} &(7.29a)\\ s.t.\ &\Gamma_1(\rho_k^l,\rho_{m,k}^u)+\chi_{1k}\le t_{\max}, &(7.29b)\\ &\Phi(\rho_k^l,\rho_{m,k}^u)\le\varrho_{m,k}, &(7.29c)\end{aligned}$$

where $\chi_{1k}\triangleq\sum_{m\in\mathcal{M}}\beta_{m,k}y_k^{l,o}/f_k^l+\beta_{m,k}y_m^{u,o}/f_{m,k}^u$, and $\Phi(\rho_k^l,\rho_{m,k}^u)$ represents the first-order Taylor approximation of $\rho_k^l\rho_{m,k}^u$, detailed as follows:

$$\begin{aligned}\Phi(\rho_k^l,\rho_{m,k}^u)\triangleq\frac{1}{4}\Big[&(\rho_k^l+\rho_{m,k}^u)^2-\big((\rho_k^l)^{(\tau)}-(\rho_{m,k}^u)^{(\tau)}\big)^2\\ &-2\big((\rho_k^l)^{(\tau)}-(\rho_{m,k}^u)^{(\tau)}\big)\big(\rho_k^l-\rho_{m,k}^u-(\rho_k^l)^{(\tau)}+(\rho_{m,k}^u)^{(\tau)}\big)\Big].\end{aligned}$$

The functions $\Gamma_1(\rho_k^l,\rho_{m,k}^u)$ and $\Gamma_2(\rho_k^l,\rho_{m,k}^u)$ are defined as follows:

$$\begin{aligned}\Gamma_1(\rho_k^l,\rho_{m,k}^u)\triangleq&\sum_{m=1}^{M}\Big(\psi_{1k,m}+(\psi_{2k,m}^s+\psi_{2k,m}^f)\rho_k^l+\psi_{2k,m}^f\rho_{m,k}^u\\ &+\psi_{3k,m}+(g_k^l)^{-1}\beta_{m,k}B_{3k}\big((\rho_k^l)^{-B_{4k}}+(\rho_{m,k}^u)^{-B_{4k}}\big)\Big),\\ \Gamma_2(\rho_k^l,\rho_{m,k}^u)\triangleq&-\sum_{m=1}^{M}\beta_{m,k}\big(e^{-B_{5k}D_k\rho_k^l}-e^{-B_{5k}D_k\rho_{m,k}^u}\big),\end{aligned}$$

with auxiliary terms defined as follows for notational compactness:

$$\psi_{1k,m} \triangleq \beta_{m,k} B_{1k} \left(\rho_k^l - 1\right)^{B_{2k}} / f_k^l, \tag{7.30}$$

$$\psi_{2k,m}^s \triangleq \beta_{m,k} D_k / r_{m,k}^s, \tag{7.31}$$

$$\psi_{2k,m}^f \triangleq \beta_{m,k} D_k / r_{m,0}^f, \tag{7.32}$$

$$\psi_{3k,m} \triangleq \beta_{m,k} C_{1m} (\rho_{m,k}^u - 1)^{C_{2m}} / f_{m,k}^u. \tag{7.33}$$

After relaxation and approximation, both the objective and constraints in (7.29) become convex, allowing efficient solution via standard convex optimization techniques.

Next, with the semantic extraction depths $\boldsymbol{\rho} \triangleq \{\rho_k^l, \rho_m^u\}_{k\in\mathcal{K},m\in\mathcal{M}}$, we turn to optimizing the UAVs' deployment. To handle non-convexities in the rate expressions, we introduce slack variables $\boldsymbol{\eta} \triangleq \{\eta_{m,k}, \eta_{m,0}\}_{m\in\mathcal{M},k\in\mathcal{K}}$ and denote the UAVs' previous positions as $\boldsymbol{\ell}_m^{old} \triangleq \boldsymbol{\ell}_m(i-1)$. The deployment subproblem is then approximated as:

$$\min_{\boldsymbol{\ell}} \sum_{k\in\mathcal{K}} \sum_{m\in\mathcal{M}} (Q_k + V\omega_1)\beta_{m,k} D_k \left(\frac{\rho_k^l}{\eta_{m,k}} + \frac{\rho_k^l + \rho_{m,k}^u}{\eta_{m,0}} \right) \tag{7.34a}$$

$$\text{s.t.} \sum_{m\in\mathcal{M}} \beta_{m,k} D_k \left(\frac{\rho_k^l}{\eta_{m,k}} + \frac{\rho_k^l + \rho_{m,k}^u}{\eta_{m,0}} \right) + \chi_{3k} \le t_{\max}, \tag{7.34b}$$

$$d_{\min} \le 2\left(\boldsymbol{\ell}_m^{(\tau)} - \boldsymbol{\ell}_{m'}^{(\tau)}\right)^T (\boldsymbol{\ell}_m - \boldsymbol{\ell}_{m'}) - \|\boldsymbol{\ell}_m^{(\tau)} - \boldsymbol{\ell}_{m'}^{(\tau)}\|^2, \tag{7.34c}$$

$$\|\boldsymbol{\ell}_m - \boldsymbol{\ell}_m^{old}\| \le t_{\max} v_{\max}, \tag{7.34d}$$

$$\eta_{m,k} \le E_{m,k} \text{ and } \eta_{m,0} \le E_{m,0}. \tag{7.34e}$$

For brevity, we define the following notations

$$\chi_{3k} \triangleq \sum_{m\in\mathcal{M}} \frac{\beta_{m,k}}{f_k^l}\left(y_k^{l,o} + B_{1k}(\rho_k^l - 1)^{B_{2k}}\right) + \frac{\beta_{m,k}}{f_{m,k}^u}\left(y_m^{u,o} + C_{1m}(\rho_{m,k}^u - 1)^{C_{2m}}\right) + \frac{\beta_{m,k} B_{3k}}{g_k^l}\left((\rho_k^l)^{-B_{4k}} + (\rho_{m,k}^u)^{-B_{4k}}\right),$$

$$E_{m,k} \triangleq \frac{-\log_2(e)\gamma_{m,k}^{(\tau)}}{(r_{m,k}^s)^{(\tau)}\|\boldsymbol{\ell}_m^{(\tau)} - \boldsymbol{q}_k\|^2}(\|\boldsymbol{\ell}_m - \boldsymbol{q}_k\|^2 - \|\boldsymbol{\ell}_m^{(\tau)} - \boldsymbol{q}_k\|^2) + (r_{m,k}^s)^{(\tau)},$$

$$E_{m,0} \triangleq \frac{-\log_2(e)\gamma_{m,0}^{(\tau)}}{(r_{m,0}^f)^{(\tau)}\|\boldsymbol{\ell}_m^{(\tau)} - \boldsymbol{q}_0\|^2}(\|\boldsymbol{\ell}_m - \boldsymbol{q}_0\|^2 - \|\boldsymbol{\ell}_m^{(\tau)} - \boldsymbol{q}_0\|^2) + (r_{m,0}^f)^{(\tau)}.$$

By applying the SCA method, the non-convex components are linearized around the current operating point, transforming the deployment subproblem into a convex form (7.34). This enables iterative refinement of UAVs' positions. The overall model-based optimization alternates between solving the semantic extraction and UAVs' deployment subproblems. Starting from the UAVs' fixed locations, semantic extraction depths are optimized based on local computational capabilities. Then, using the updated extraction strategy, UAVs' positions are adjusted via SCA to improve communication efficiency. This alternating process continues until convergence, yielding a jointly optimized solution for UAVs' deployment and semantic extraction that effectively reduces the SAoI across all GUs.

7.5 Numerical Evaluation

This section presents simulation results to evaluate the effectiveness of the proposed Hily-PPO framework. The considered network consists of 3 UAVs and 5 GUs, with the BS positioned at (900, 500) in meters. The GUs are randomly deployed within a 1000×1000 m^2 rectangular area on the (x, y)-coordinate. Key simulation parameters are set as follows: maximum UAV speed $v_{\max} = 30$ m/s, minimum inter-UAV distance $d_{\min} = 30$ m, transmission power for both GUs and UAVs $p_k = p_u = 35$ dBm, and the weighting factor $V = 100$. For performance comparison, several benchmark schemes are implemented including Max-AoI, Max-Value, and Non-Extraction (Non-Ext) schemes. In the Max-AoI scheme, each UAV selects the GU with the highest current AoI within its communication range and proceeds to collect data from that GU. The Max-Value strategy prioritizes access based on information significance, allowing the GU with the highest data value to utilize the uplink UAV-GU channel for transmission. In contrast, the Non-Ext scheme omits semantic processing entirely, where GUs send unprocessed raw data to the BS through multiple UAV relays. Since no information is discarded during transmission, this approach achieves the maximum possible information value at the BS and thus serves as a performance upper bound in terms of received information quality.

7.5.1 Learning Efficiency and Convergence

Figures 7.4a and 7.4b illustrate the convergence behavior of the model-free PPO and the model-based AO approach, respectively, within a single time slot. In Fig. 7.4a, the hierarchical PPO is compared against the conventional model-free PPO, where the latter jointly optimizes all the variables $\{\beta_{m,k}, \ell_m, \rho_k^l, \rho_{m,k}^u\}_{k\in\mathcal{K}, m\in\mathcal{M}}$ in a monolithic action space. Due to the high dimensionality and mixed discrete-continuous nature of this space, the conventional PPO suffers from slow and unstable convergence. In contrast, the hierarchical PPO leverages guidance from the model-based method to decompose the decision process, effectively reducing the complexity of the policy search. This structural advantage enables the proposed algorithm to achieve both faster convergence and superior reward performance. Figure 7.4b demonstrates the iterative convergence of the model-based AO optimization. The reward drops sharply

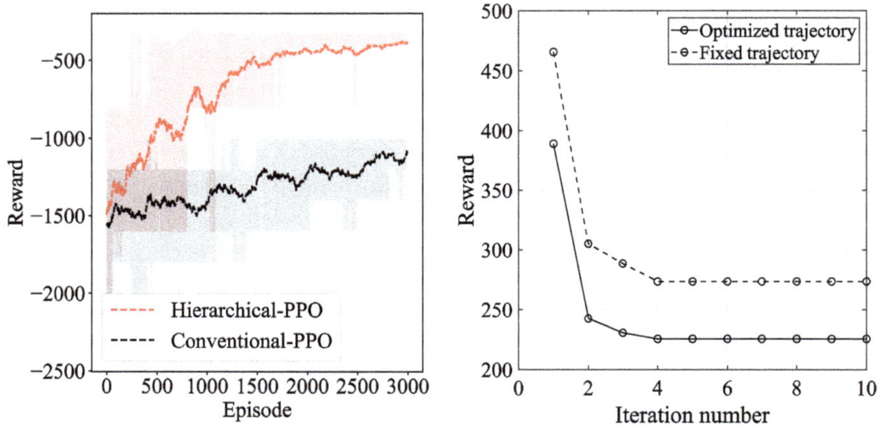

(a) PPO for UAV-GU association (b) Model-based AO method

Fig. 7.4 **a** Hierarchical PPO achieves faster, more stable convergence than conventional PPO by reducing action space with model-based guidance. **b** The model-based AO method converges rapidly within a few iterations, as UAVs efficiently optimize their positions and semantic extraction policies

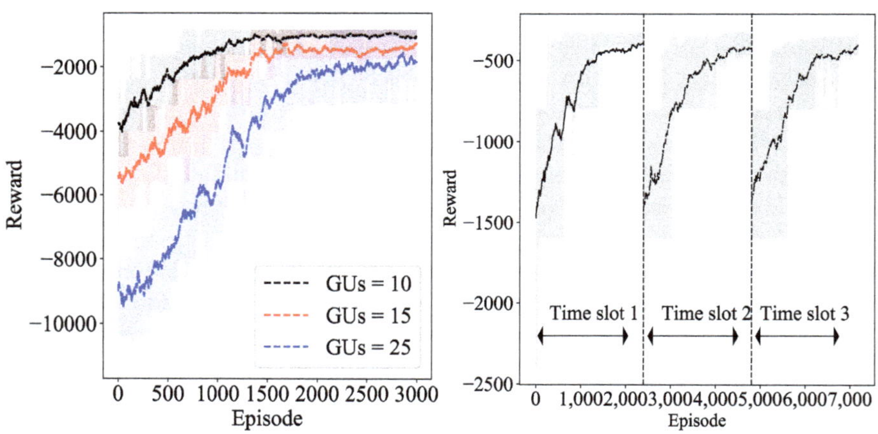

(a) PPO's convergence in a single slot. (b) Convergence across successive slots.

Fig. 7.5 **a** Hily-PPO's convergence becomes slow with more GUs due to increased action space complexity. **b** In each time slot, model-based AO provides initial control action and then model-free PPO refines them, with rewards improving as policies stabilize

in the initial iterations and stabilizes within a small number of steps, indicating low computational overhead per time slot. During this process, each UAV adjusts its location to an optimal hovering point. As the UAVs' positions and semantic extraction policies are refined, the system reward improves rapidly before reaching a steady state, reflecting the efficiency of the model-free optimization procedure.

The convergence characteristics of the Hily-PPO framework under varying network scales are shown in Fig. 7.5a, with different numbers of GUs: $K = 10$, $K = 15$, and $K = 25$. The results indicate that convergence speed is sensitive to the number of GUs. With fewer GUs, the action space is more compact, leading to faster policy stabilization. As K increases, the dimensionality of the UAV-GU association and control space grows significantly, introducing greater exploration space and delaying convergence. Thus, more training episodes are required for the agent to learn effective strategies in dense networks. Figure 7.5b presents the multi-slot convergence of Hily-PPO with a fixed $K = 5$. At the start of each time slot i, the model-based AO method computes an initial estimate of the UAVs' positions $\boldsymbol{\ell}(i)$ and semantic extraction depths $\boldsymbol{\rho}(i)$ based on the current UAV-GU association strategy $\boldsymbol{\beta}(i)$. The UAVs then execute the joint action $\{\boldsymbol{\beta}(i), \boldsymbol{\ell}(i), \boldsymbol{\rho}(i)\}$, prompting updates in the GUs' AoI and information value. These updated states serve as the initial conditions for the next time slot. Within each slot, the reward initially decreases due to strategic adjustments in UAV-GU pairing, but subsequently rises as the model-free PPO fine-tunes the policy. Once a stable and high-performing joint action is attained, the UAVs transmit their current locations and semantic strategies to the BS for coordination and monitoring.

7.5.2 UAVs' Trajectories and AoI Dynamics

Figure 7.6 depicts the UAVs' dynamic trajectories over time under the proposed control framework, with 10 GUs uniformly spread across the operational area. Three UAVs autonomously divide the coverage region and cooperatively serve the distributed GUs. Specifically, UAV-1 is assigned to GU-7, GU-9, and GU-10. UAV-2

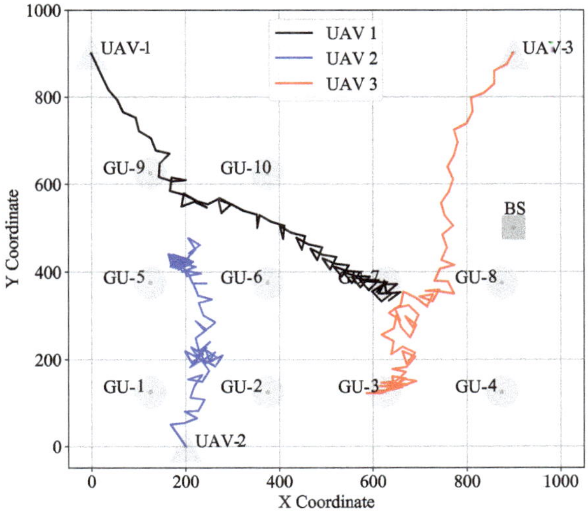

Fig. 7.6 UAVs dynamically adjust their trajectories to serve GUs

handles GU-1, GU-2, GU-5, and GU-6, while UAV-3 covers the remaining GUs. At the beginning of the operation, all UAVs navigate toward their designated service zones to minimize communication distances and improve link quality. Once in proximity, they hover to ensure reliable data exchange. The resulting trajectories reflect adaptive decision-making and spatial coordination among UAVs. The non-smooth nature of the UAVs' trajectories arises from real-time adjustments in UAV-GU associations, which are driven by the instantaneous AoI levels, GUs' locations, and the semantic value of information to be transmitted. As the system prioritizes minimizing SAoI, the association policy evolves dynamically based on changing network conditions. This leads to changes of trajectories, such as redirection or relocation, when serving new or higher-priority GUs.

Figure 7.7 shows the temporal variation of the network's average AoI and information value under various schemes. For the proposed Hily-PPO framework, both metrics initially degrade before stabilizing. The initial increase in AoI is due to the UAVs' suboptimal deployment locations. When UAVs start far from their target GUs, channel conditions are weak, leading to low transmission rates and high latency, thereby increasing AoI. As UAVs adjust their positions and move closer, communication efficiency improves significantly, causing a sharp drop in AoI. However, this improvement in freshness comes at the cost of reduced information value. This trade-off stems from the system's adaptive semantic extraction. To accelerate transmission and reduce delay, the framework extracts only essential semantic content, discarding less critical details. While this strategy effectively reduces AoI, it inherently reduces the richness of the delivered information. In comparison, Non-Ext

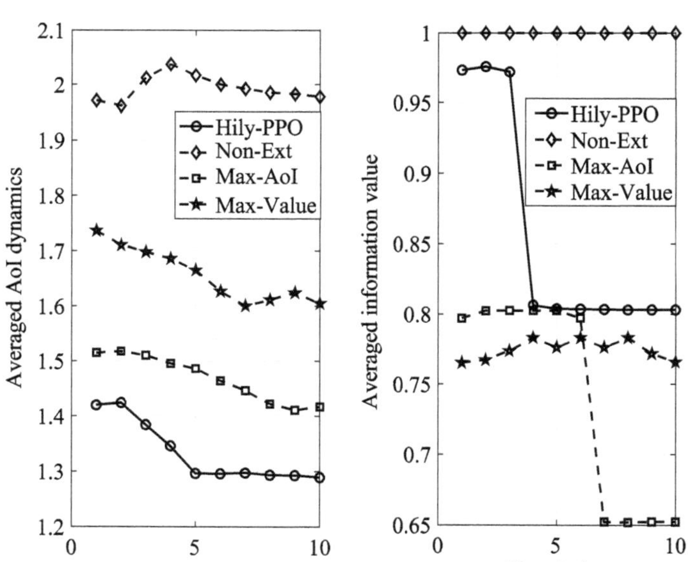

Fig. 7.7 Hily-PPO achieves the lowest AoI and maintains high value of information, while the other baselines either suffer high AoI or fail to balance freshness and value effectively

consistently exhibits high AoI throughout the simulation, highlighting its inefficiency of transmitting complete raw data without compression, which results in prolonged transmission times. Max-AoI achieves better AoI performance than Max-Value by actively selecting GUs with the most outdated information, thus preventing excessive staleness. Conversely, Max-Value focuses on maximizing information value, choosing large semantic payloads that increase transmission duration and consequently lead to higher AoI.

7.5.3 AoI and Semantic Value Trade-Off

Figure 7.8 illustrates the average AoI as a function of the data size generated by the GUs. Besides learning based methods, we also implement Lyapunov decomposition and approximate each subproblem by conventional optimization-based methods, denoted as Lya-Opt in Figs. 7.8 and 7.9. The proposed Hily-PPO framework consistently achieves the lowest AoI across all data size settings, outperforming the baseline methods. This performance gain arises from its joint optimization of semantic extraction depth and UAVs' trajectory positions. As GUs' data size increases, both uplink transmission time and propagation delay rise, leading to a higher AoI. However, Hily-PPO counteracts this effect by intelligently optimizing UAV-GU associations and UAVs' trajectories to minimize communication latency. With large data sizes, Max-Value surpasses Max-AoI in AoI performance. This shift highlights the growing significance of prioritizing high-value semantic content when network resources are strained by bulky data transmissions. For large payloads, extracting meaningful semantics demands substantial processing, introducing non-negligible semantic encoding delays. If not carefully managed, such overhead can degrade timeliness. Max-Value focuses on information richness and becomes more beneficial under these conditions. Figure 7.8 also displays the information value achieved by different schemes. Larger data volumes generally yield higher information value. Among them, Hily-PPO not only delivers the highest information value but also closely approaches the upper limit. This is achieved through joint control of AoI and semantic fidelity, ensuring that timely updates do not come at the expense of excessive semantic loss.

We examine the average AoI and information value as the number of GUs increases, as shown in Fig. 7.9. The SAoI metric, defined as a weighted combination of AoI and information value in (7.17), reflects the overall freshness-quality trade-off. Hily-PPO maintains superior performance in both AoI and SAoI compared to the baseline schemes. As the number of GUs grows, competition for uplink access intensifies due to the UAVs' limited service capacity. This congestion leads to longer update intervals and consequently higher AoI. Moreover, the GUs' average information value tends to decline as the number of GUs increases. Hily-PPO still maintains a higher information value by dynamically allocating semantic processing and wireless resources, emphasizing transmissions with high semantic importance.

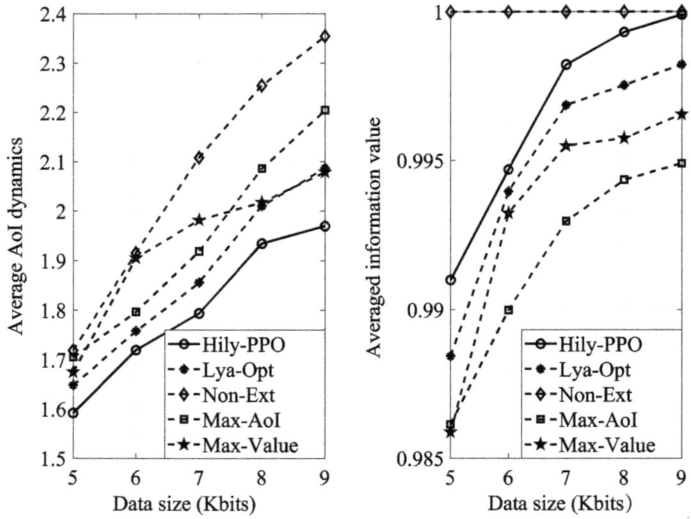

Fig. 7.8 Hily-PPO achieves the lowest AoI and highest information value across varying data sizes by jointly optimizing UAVs' trajectories, UAV-GU associations, and semantic extraction. Under large payloads, Max-Value outperforms Max-AoI as prioritizing semantic value becomes more effective. Hily-PPO balances timeliness and semantic fidelity, maintaining near-optimal information value while minimizing AoI, demonstrating superior adaptability to network load variations

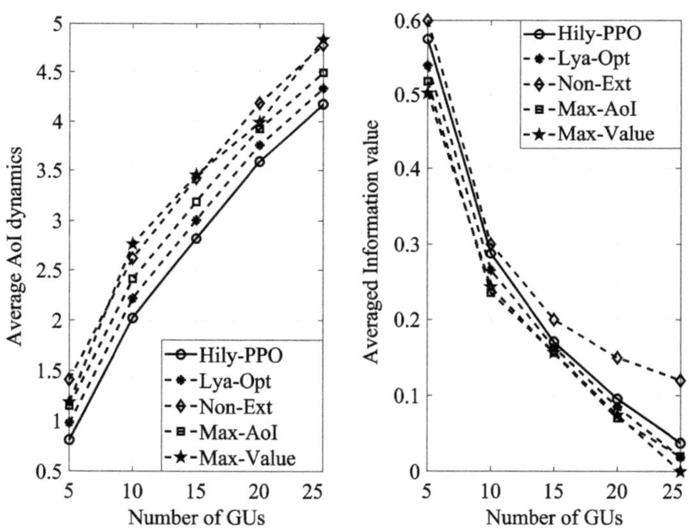

Fig. 7.9 As the number of GUs increases, network congestion raises AoI and reduces average information value. Hily-PPO outperforms baselines in both AoI and information value by dynamically prioritizing high-value semantic transmissions and efficiently allocating resources, maintaining superior freshness and quality under growing user density and limited service capacity

7.6 Discussions and Conclusions

This case study investigates the critical trade-off between data freshness in terms of AoI and the semantic value of transmitted information in a UAV-assisted wireless network. Recognizing that simply minimizing latency is insufficient when the meaningful content of data is paramount, the study formulates a time-averaged semantic-aware AoI (SAoI) minimization problem, which integrates both temporal freshness and semantic relevance. The Hily-PPO framework was introduced to solve this stochastic optimization problem by combining Lyapunov optimization with a hierarchical PPO algorithm. First, Lyapunov optimization decomposes the long-term SAoI minimization into a series of per-slot subproblems, enabling online decision-making without requiring future network knowledge. Then, for each subproblem, a hierarchical PPO algorithm is devised to determine the optimal actions efficiently, by leveraging the model-free PPO to adapt the UAV-GU association policies in response to dynamic network states. Concurrently, the semantic data extraction strategy and UAVs' trajectories are optimized using a model-based iterative optimization method, which leverages known system information for precise control. This hierarchical algorithm balances the flexibility of model-free learning with the efficiency of model-based optimization. Extensive simulations demonstrate that the Hily-PPO framework significantly improves data's freshness while consistently preserving high semantic information value under various network scenarios, outperforming conventional schemes that focus solely on AoI or throughput.

More importantly, this case study demonstrates that model-free DRL can be efficiently integrated into the Lyapunov optimization framework for solving stochastic resource allocation problems under dynamic network conditions. As discussed in previous chapters, the conventional model-based optimization methods can be used to guide the model-free DRL methods to improve learning efficiency and reward performance. Conversely, our case study in this chapter bridges the gap in the opposite direction by showing that the model-free DRL methods can also drive the stochastic optimization to approximate the optimal solutions. Such a bidirectional synergy opens new avenues for integrated algorithm design. Following this research direction, our future work will further explore the interactive coupling, mutual benefit, and integrated algorithm designs between model-based optimization methods and model-free learning methods for solving complex network control problems.

References

1. Y. Long, S. Gong, S. Sun, G.C.F. Lee, L. Li, D. Niyato, Lyapunov-guided deep reinforcement learning for semantic-aware AoI minimization in uav-assisted wireless networks. IEEE Trans. Wireless Commun. **24**(8), 6351–6364 (2025)
2. R.D. Yates, Y. Sun, D.R. Brown, S.K. Kaul, E. Modiano, S. Ulukus, Age of information: an introduction and survey. IEEE J. Sel. Areas Commun. **39**(5), 1183–1210 (2021)
3. I. Kahraman, A. Köse, M. Koca, and E. Anarım, Age of information in internet of things: a survey. IEEE Internet Things J. **11**(6), 9896–9914 (2023)

4. A. Kosta, N. Pappas, V. Angelakis, Age of information: a new concept, metric, and tool. Found. Trends® Netw. **12**(3), 162–259 (2017)
5. A. Li, S. Wu, J. Jiao, N. Zhang, Q. Zhang, Age of information with Hybrid-ARQ: a unified explicit result. IEEE Trans. Commun. **70**(12), 7899–7914 (2022)
6. S. Wu, Z. Deng, A. Li, J. Jiao, N. Zhang, Q. Zhang, Minimizing age-of-information in HARQ-CC aided NOMA systems. IEEE Trans. Wireless Commun. **22**(2), 1072–1086 (2023)
7. M. Xi, H. Dai, J. He, W. Li, J. Wen, S. Xiao, and J. Yang, A lightweight reinforcement learning-based real-time path planning method for unmanned aerial vehicles. IEEE Internet Things J. **11**(12), pp. 21061–21071 (2024)
8. R. Liu, M. Xie, A. Liu, H. Song, Joint optimization risk factor and energy consumption in IoT networks with tinyML-enabled internet of UAVs. IEEE Internet Things J. **11**(12), 20983–20994 (2024)
9. Y. Emami, H. Gao, K. Li, L. Almeida, E. Tovar, Z. Han, Age of information minimization using multi-agent UAVs based on AI-enhanced mean field resource allocation. IEEE Trans. Veh. Technol. **73**(9), 13368–13380 (2024)
10. J. Liu, F. Yang, X. Wang, L. Qu, M. Jin, H. Dai, Joint optimization of charging station placement and UAV trajectory for fresh data collection. IEEE Internet Things J. **11**(14), 25057–25073 (2024)
11. Y. Long, S. Zhao, S. Gong, B. Gu, D. Niyato, X.S. Shen, AoI-aware sensing scheduling and trajectory optimization for multi-UAV-assisted wireless backscatter networks. IEEE Trans. Veh. Technol. **73**(10), 15440–15455 (2024)
12. B. Zhu, E. Bedeer, H.H. Nguyen, R. Barton, Z. Gao, UAV trajectory planning for AoI-minimal data collection in UAV-aided IoT networks by transformer. IEEE Trans. Wireless Commun. **22**(2), 1343–1358 (2023)
13. W. Qi, C. Yang, Q. Song, Y. Guan, L. Guo, A. Jamalipour, Minimizing age of information for hybrid UAV-RIS-assisted vehicular networks. IEEE Internet Things J. **11**(10), 17886–17895 (2024)
14. B. Chen, D. Liu, J. Zhang, L. Hanzo, Learning-aided UAV-cooperation reduces the age-of-information in wireless networks. IEEE Commun. Lett. **28**(5), 1053–1057 (2024)
15. M. Samir, C. Assi, S. Sharafeddine, A. Ghrayeb, Online altitude control and scheduling policy for minimizing AoI in UAV-assisted IoT wireless networks. IEEE Trans. Mobile Comput. **21**(7), 2493–2505 (2022)
16. Q. Sun, J. Niu, X. Zhou, T. Jin, Y. Li, AoI and data rate optimization in aerial IRS-assisted IoT networks. IEEE Internet Things J. **11**(4), 6481–6493 (2024)
17. H. Xie, Z. Qin, G.Y. Li, B.-H. Juang, Deep learning enabled semantic communication systems. IEEE Trans. Signal Process. **69**, 2663–2675 (2021)
18. Z.Q. Liew, M. Xu, W.Y.B. Lim, Z. Xiong, D. Niyato, D.I. Kim, Mechanism design for semantic communication in UAV-assisted metaverse: a combinatorial auction approach. IEEE Trans. Veh. Technol. **73**(2), 2236–2251 (2024)
19. X. Kang, B. Song, J. Guo, Z. Qin, F.R. Yu, Task-oriented image transmission for scene classification in unmanned aerial systems. IEEE Trans. Commun. **70**(8), 5181–5192 (2022)
20. P. Yue, J. Xin, Y. Zhang, Y. Lu, M. Shan, Semantic-driven autonomous visual navigation for unmanned aerial vehicles. IEEE Trans. Ind. Electron. **71**(11), 14853–14863 (2024)
21. J. Kang, H. Du, Z. Li, Z. Xiong, S. Ma, D. Niyato, Y. Li, Personalized saliency in task-oriented semantic communications: image transmission and performance analysis. IEEE J. Sel. Area. Commun. **41**(1), 186–201 (2023)
22. L. Wang, W. Wu, F. Tian, H. Hu, Intelligent resource allocation for UAV-enabled spectrum sharing semantic communication networks, in Proceedings of IEEE International *Conference on Communication Technology (ICCT)*, pp. 1359–1363 (2023)
23. J. Chen, S. Yang, T.-T. Chan, H. Pan, Enhancing information freshness via knowledge graph-aided semantic communication, in Proceedings of *IEEE International Conference on Information, Communication and Networks (ICICN)*, pp. 155–160 (2023)
24. Z.Q. Liew, M. Xu, W.Y.B. Lim, Z. Xiong, D. Niyato, D.I. Kim, Mechanism design for semantic communication in UAV-assisted metaverse: a combinatorial auction approach. IEEE Trans. Veh. Technol. **73**(2), 2236–2251 (2024)

25. S. Meng, S. Wu, A. Li, Q. Zhang, Toward goal-oriented semantic communications: AoII analysis of coded status update system under FBL regime. IEEE J. Sel. Area. Inf. Theory **4**, 718–733 (2023)
26. Q. Wu, Y. Zeng, R. Zhang, Joint trajectory and communication design for multi-UAV enabled wireless networks. IEEE Trans. Wireless Commun. **17**(3), 2109–2121 (2018)
27. G. Zhang, Q. Wu, M. Cui, R. Zhang, Securing UAV communications via joint trajectory and power control. IEEE Trans. Wireless Commun. **18**(2), 1376–1389 (2019)
28. Z. Yang, M. Chen, Z. Zhang, C. Huang, Energy efficient semantic communication over wireless networks with rate splitting. IEEE J. Sel. Areas Commun. **41**(5), 1484–1495 (2023)
29. Z. Wang, J. Zhao, Utility-oriented wireless communications for 6G networks: semantic information transfer for IRS aided vehicular metaverse, in Proceedings of *IEEE Vehicular Technology Conference (VTC)* (Florence, Italy, Jun, IEEE VTC-Spring, 2023), pp. 1–7
30. M. Neely, Energy optimal control for time-varying wireless networks. IEEE Trans. Inf. Theory **52**(7), 2915–2934 (2006)
31. J. Schulman, F. Wolski, P. Dhariwal, A. Radford, O. Klimov, Proximal policy optimization algorithms. arXiv:1707.06347 (2017). [Online]. Available: http://arxiv.org/abs/1707.06347

Chapter 8
Summary

8.1 Summary of Contributions

This book has presented a comprehensive exploration of the integration between conventional model-based optimization and modern model-free DRL methods for future wireless networks. Through case studies and detailed analysis, we have demonstrated how this addresses the fundamental challenges of resource allocation and performance optimization in increasingly complicated, decentralized, heterogeneous, and dynamic wireless networks.

Our investigation began with a review of RL/DRL fundamentals and the analysis of their design principles, applicability, and computational characteristics, highlighting their robustness, flexibility, and generic frameworks for typical wireless networks. We also highlighted both strengths and weaknesses of conventional model-based optimization methods, which leverage well-established prior knowledge to achieve high efficiency and even provable performance guarantees, but also struggle to cope with the scalability, uncertainty, and complexity of future wireless networks. We then systematically examined several innovative frameworks and integrated algorithm designs to exploit the interplay between model-based optimization and model-free DRL. In particular, our main technical contributions are revealed through detailed case studies in different chapters and summarized as follows:

- Optimization-driven DRL was proposed as a powerful paradigm that leverages partial model information to guide and accelerate the learning process. By incorporating domain knowledge through approximate optimization formulations, this approach significantly reduces training overhead while improving convergence performance. The case study in Chap. 2 focused on high dimensional beamforming optimization in an IRS-assisted wireless network, including the RF transmitter's active beamforming and the IRS's passive beamforming strategies. The active beamforming optimization has been extensively studied in literature, even with closed-form solutions in simple cases, which can serve as the domain knowledge to expedite the search for passive beamforming by model-free DRL.

S. Gong et al., *Optimization-Driven Deep Reinforcement Learning for Wireless Networks*, Wireless Networks, https://doi.org/10.1007/978-3-032-22997-7_8

- Hierarchical DRL was discussed as indispensable in heterogeneous wireless networks for decomposing complex network control problems into manageable subtasks at individual network entities. This division allows for efficient coordination between model-free DRL in outer loops and model-based optimization in inner loops, enabling scalable solutions for large-scale networks. The case study in Chap. 3 focused on throughput maximization in a backscatter-aided relay communications system by jointly optimizing the relay selection and transmission control strategies. The combinatorial relay selection is iteratively updated by the outer-loop DDPG, while the beamforming and time allocation strategy are determined by the inner-loop optimization. In Chap. 4, a similar hierarchical DRL was applied to AoI minimization problem in an IRS-assisted wireless network. The outer-loop PPO algorithm was used to determine the transmission scheduling decision in each time slot. Given the outer-loop scheduling decision, the inner-loop optimization updates the transmission control parameters more efficiently. Such a hierarchical PPO method was shown to achieve a significantly higher reward performance and faster convergence.
- Hierarchical MADRL was considered as a promising technique for coordinated control in decentralized and heterogeneous wireless networks. The hierarchical MADRL in Chap. 5 includes the high-level agent at the AP to update its beamforming strategy and the low-level user agents to learn and adapt their offloading strategies. At the high-level agent with more powerful capabilities for information collection and computation, model-based optimization can be implemented to improve the learning efficiency by providing action estimates for low-level user agents as better-informed starting points for action exploration. Each user agent can further update its own action by model-free DRL, assuming that all other agents follow the model-based action estimation. This can parallelize MADRL into multiple single-agent DRL, allowing each agent to update actions independently based on local observations. As such, the state and action spaces can be reduced dramatically to improve the learning efficiency.
- Bayesian-enhanced DRL was proposed as a robust learning framework to handle information uncertainty and limited communications in large-scale decentralized wireless networks. By integrating Bayesian inference with DRL, we demonstrated improved adaptability to time-varying channel conditions and measurement uncertainties. The case study in Chap. 6 focused on a UAV-assisted wireless network for GUs to offload their sensing data to remote BS. MADRL was used to adapt the UAVs' trajectory planning and network formation strategies to minimize the system energy consumption and transmission delay. Considering limited communications among UAVs, Bayesian optimization was further employed for each UAV to estimate a more preferable hovering location for data collection based on historical trajectory points. This helps avoid inefficient action explorations and improves the convergence performance.
- Lyapunov-guided DRL demonstrated that the model-free DRL methods can also be efficiently integrated into optimization framework, especially the multi-slot stochastic optimization for solving the long-term resource allocation problems under dynamic network conditions. It can effectively decompose complex stochas-

tic optimization into tractable per-slot subproblems in multiple time slots. Model-free DRL can be further invoked to approximate each subproblem in each time slot and thus drive the temporal evolution of the system, ensuring asymptotic achievement of the stochastic objective and satisfaction of the long-term resource constraints. The case study in Chap. 7 demonstrated that it can effectively solve the stochastic minimization problem in a UAV-assisted semantic communication network and verified that it can improve information freshness and preserve semantic integrity under varying network conditions.

8.2 Future Research Directions

Several important challenges and opportunities remain open for future investigation. The current challenges and open problems can be broadly categorized into the following aspects. First, it is challenging to develop more sophisticated and lightweight hybrid frameworks that can balance the use of model-based optimization and model-free DRL. This may require adaptive mechanisms to dynamically switch between model-based optimization and model-free DRL based on real-time network conditions and resource budgets. Second, it is unsatisfactory to verify the practical effectiveness of the hybrid frameworks merely based on limited experiments or field tests. It is imperative to solidify the theoretical foundation of these hybrid algorithm frameworks by thorough analysis and rigorous proof for convergence, stability, and optimality. Another challenge is the scalability of these integrated algorithm designs to ultra-dense and large-scale heterogeneous networks, which requires further exploration in terms of distributed execution, communication overhead, and convergence guarantees. Finally, it is still challenging to bridge the gap between theoretical analysis and practical deployment, urging more research on real-world validation and standardized testing environments. Addressing these challenges will be crucial for advancing the integration of model-based optimization and model-free DRL towards realizing scalable, efficient, and intelligent wireless networks.

To conclude this book, we identify a few promising research problems in this topic to invoke more insightful ideas.

- *Communication-limited MADRL* can be commonly observed in decentralized wireless networks that make it very difficult for timely information sharing between different agents. Conventional MADRL relies on centralized training and decentralized execution. This requires complete information of different network entities during the training phase. However, the real-time information exchange among UAVs in a dynamic wireless network is usually unreliable or intermittent due to the stochastic nature of wireless communications. The UAVs serving different regions can be deployed far away and thus have no direct communication links with each other and the base station. As such, the information collected from different UAVs can be incomplete, asynchronous, and perturbed by random noise and time delay. This calls for delay-tolerant MADRL to enable efficient learning with incomplete information in decentralized wireless networks. A straightforward intuition is that more performance loss will be observed with less information available for centralized training. Hence, a simple implementation of the delay-tolerant MADRL is to penalize the UAVs' information delay and encourage the UAVs' frequent information exchange on their trajectories. Specifically, the information delay incurred penalty term in the reward function can drive different UAVs to build regular or periodical communication links with each other. As such, MADRL can be still applicable and effective with little performance loss.

- *Compensation for the information loss* can be possible by exploring the spatio-temporal correlations for information estimation and prediction. To dealing with information loss in UAV-assisted wireless networks, a more proactive approach is to estimate the missing information by machine learning approaches based on the UAVs' past behaviors and information exchange with the base station. The UAV may have sporadic information exchange when it is close to the base station along its trajectory. The base station can record the UAV's historical information and use it to predict the UAV's future status. Besides the temporal prediction, different UAVs may also have hidden spatial correlations. For example, two UAVs can be highly correlated if they have similar trajectories or observe similar users' traffic demands. Such spatio-temporal correlations can be jointly exploited to improve the prediction performance by using advanced machine learning techniques, e.g., graph neural network and attention mechanism.
- *Networking for learning* will be a more preferable option for network design and deployment to meet the demand for frequent information exchange and efficient computation during the implementation of machine learning approaches in future wireless networks. Conventionally, the deployment of UAV-assisted wireless networks mainly aims to extend network coverage, reduce transmission latency, improve network capacity, and energy efficiency, etc., generally focusing on the communication and networking perspective. As the future wireless networks become more heterogeneous, decentralized, and complicated, the network design problems will largely rely on modern machine learning approaches. In particular, the successful deployment of MADRL requires timely information exchange among different UAVs. This implies that the MADRL-driven UAV-assisted wireless networks should provide more communication opportunities for different UAVs to exchange local information. However, this can be contradictory with the design target from the networking perspective. For example, to maximize the network capacity, each UAV will be assigned a separate service region with minimum overlap. This will result in limited communications among the UAVs that may prevent the MADRL's successful deployment. The concept of networking for learning motivates us to focus more on the practical implementation of machine learning approaches during the network design, finding a balance between the design requirements from both machine learning and networking perspectives.
- *Simultaneous learning and modeling (SLAM)* can be realized in an evolving digital twin to expedite DRL with limited interactions with the real wireless network environment. The implementation of DRL approaches is basically a continuous trial-and-error process by interacting with the environment. This can be feasible

when frequent interaction is efficient with low cost and the reward can be easily observed. However, in practical wireless networks, the reward of each action taken to the network is not obviously observable and the interactions with all other network entities demand considerable time overhead and resource consumption. For example, in UAV-assisted wireless networks with limited energy supply, it becomes intolerable for the UAV to freely explore the space before planning its optimal trajectory. Without trial-and-error in the real network environment, the implementation of DRL approaches can be problematic, which motivates us to construct a digitized virtual twin model for the real environment to avoid costly interactions. In particular, we can deploy machine learning approaches in the digital twin and implement a mature policy in the real environment after numerous training in the digital twin. Moreover, the digital twin model can be dynamically evolving with the learning process instead of an offline pre-trained model. Specifically, given an initial twin model, we can plan an initial policy for the network by machine learning approaches. When the initial policy is practically implemented in the real network environment, we can observe the network's feedback and record new data samples to update the digital twin model. The updated digital twin in return will provide a more accurate virtual environment for planning a better policy for the network in the next step. Such a simultaneous learning and modeling approach can be viewed as an extension of the traditional simultaneous localization and mapping concept from the physical world to the cyber-logical domain.

Through extensive case studies, we have demonstrated the complementary nature and mutual benefits achievable by strategic integration of model-based optimization and model-free DRL within a unified framework. This represents a paradigm shift for more efficient and intelligent network control, unlocking new possibilities for solving complex network optimization problems at scales previously unattainable. The future research directions highlighted above are not exhaustive but are intended to spark new ideas and drive future innovation. The widespread adoption and successful deployment of these techniques depend on consistent collaboration among wireless network researchers, AI scientists, and industry practitioners. We hope this book serves as both a comprehensive reference and a source of inspiration for researchers and engineering students to explore and innovate at the intersection of AI and wireless communications.

www.ingramcontent.com/pod-product-compliance
Ingram Content Group UK Ltd.
Pitfield, Milton Keynes, MK11 3LW, UK
UKHW020934180926
475692UK00003B/33
* 9 7 8 3 0 3 2 2 2 9 9 6 0 *